FOURTEEN DAYS
REFLECTIONS ON LIFE,
MIND, AND OTHER THINGS

FOURTEEN DAYS

Reflections on Life, Mind, and other Things

Antonio Machado Carrillo

Fourteen days
Reflections on life, mind, and other things
Author: Antonio Machado Carrillo

Spanish 1st edition February 2018
Catorce días. Reflexiones sobre la vida, la mente y más cosas
CALIGRAMA, 214 pp.

English edition: February 2019
Translation: Ana Lima
Independent publication
ISBN: 9781794628601

Copyright © 2019 Antonio Machado Carrillo
All rights reserved

* * *

This is a good opportunity to explain to English-speaking people that names in Spanish are formally constructed differently from English in order to keep track of the genealogy, of who is descendant of whom. It is the first surname that is mostly used and even if we write our name with both surnames, the title is only followed by the first one. In my case, it reads:

Christian name (Antonio) + Father's family name (Machado) + Mother's family name (Carrillo) = Antonio Machado Carrillo. Thus, I am Mr. Machado not Mr. Carrillo.

*This book is dedicated to you,
and to nobody else*

Preface

TODAY IS MONDAY, 24 JULY 2017. I've finally found a nice desk; I have plenty of paper, a new ink bottle, my loyal Pelikan, mountain coffee, my pipe filled with Virginia tobacco, peace and quiet, and time ahead. This time in the afternoon is most peaceful.

I can hear birds hidden by the thick of the forest, the chirping of insects—I'll eventually find out which—and the ceiling fan flapping, making more bearable the humid heat that will make me sweat for a couple of days, until my physiology gets used to the climate change. I've sought to be away from my island, my family and professional matters, as all of them take up most of my time and only leave a few crumbs of peace and quiet at the end of the day, which are not enough for the task I've taken upon myself.

Rather than a task, it's a promise I made to my wife a couple of years ago that seemed would never be fulfilled. It came about innocently, one ordinary day when we were discussing life and people, their desires and concerns. And in the midst of the chatter that was becoming some kind of a debate, she appeared to be surprised at how easily I talked about my ideas on the meaning of life; my clear understanding of issues that have been so widely discussed throughout history. We reached a point when we both fell into a thoughtful silence, something odd in her as she's a woman of character and untameable when discussing; it goes with her profession as a journalist, I guess.

"Well, if it's so obvious to you...write it!", she defied me. I took up the challenge out of gallantry or perhaps because I sensed that it might not be totally crazy to put in writing my particular view on "life". Having accepted the challenge, I've often been over the issue, especially when I'm having a shower, as I'm a shower thinker, but I must confess all this morning creativity usually goes down the drain, as if the water jealously grabbed my ideas, slide them down my body and carry them towards the sea.

The idea of wishing to offer something new on such a hackneyed topic as the meaning of life does sound pretentious, but that's not where I'm aiming. I must warn you I have no intention of entering into the ring of philosophical, metaphysical or religious essays which no doubt fill shelves and shelves. In fact, I think that since I was a teenager and after realizing I'm an atheist, I've read no one in particular. And I wouldn't do a good job in confronting my ideas with those of thinkers specializing in the matter, if there was something to be discussed after all.

I'm a scientist, and as such, I'm familiarised with the formalities imposed by Science on the way knowledge is generated. I'll talk about this at some stage. On occasion, I've approached human issues from a perhaps, socio-biological perspective—to label it somehow—following the rules of essays, but in a rather unorthodox manner; almost like an intellectual amusement. But this is not the case now. I renounce *ab initio* to the intention of proving something and I won't discuss other people's ideas either. With this text, personal and subjective, I want to convey the way I see certain things and understand the world, no more. Proselytism appals me and I do hope not to fall in its trap. If I do, please be indulgent and let this warning be your antidote.

To start off, I must make it clear that when I talk about life I mean both the natural phenomenon studied by Biology and our life, the life of human beings, which is so different from that of other animal species, or an individual's personal life, with its lights and shadows. These three "lives" are related and the nice thing about it is perhaps the way in which this writer has come to understand the life of human beings as a very peculiar life-mind amalgam, which sets me off into a deep explanatory mission. I stress "explanatory" because I only mean to make you see why I understand things the way I do, whether you share my views or not.

To avoid any suspicion that could sneak into these explanations I've adopted a special format for this text, writing it as a travel notebook,—reflective, true—but also personal and subjective as the genre demands. That's why I've resorted to fountain pen and paper as the best means to convey my ideas "*a capella*"—if I may say so—entwined with my experiences and trifles in this 14-day holiday I've given myself. Later, with time, when I'm back on my island of Tenerife, I'll make comments on certain points by means of notes at the end of the text. This way I'll avoid the branches being heavier than the trunk.

Well! I've ended up introducing something like a method, as used in scientific papers. Apparently, it's true a leopard can't change its spots.

I've been interrupted by Tong, the young Laotian girl who brings more steaming coffee, with that broad smile so common in the Asian southeast. I haven't yet said that Laos is my chosen destination for this break. I'm some twelve kilometres from Luang Prabang, in a small resort made of bungalows embedded in the tropical forest that lines one of the banks of the Namkhan River, tributary of the Mekong, and quite removed from everything.

It's called Zen Namkhan Boutique Resort and it's known locally as *The Resort* but I'll call it Zen Namkhan, not to give any extra points to the perfidious Albion[1].

I chose to come in July and August, in the height of the monsoon, hoping that the rain would drive away tourists, who are increasingly infesting this part of Asia. So far, so good. There's only a couple from New Zealand, both in their late middle-age (like myself), eating at a table not very close to mine in this dining-room pagoda, which opens to the surrounding forest. I notice they're the type of people who scoop like wading birds to bring their mouths to the spoon while they keep their forearm nailed on the edge of the table. When desert arrives, they each bring out their modern mobile phones and start to fiddle with them, implementing an interpersonal "brexit" that is rather pitiful to watch.

I focus on my dish since a combination of aesthetics and psychology can very well ruin a good meal. I've just been served a local vegetable tempura which is crunchy and extraordinarily tasty. I have it with a modest chardonnay; it might be from New Zealand...

To make things worse, the gentleman announces the effort of removing himself from the chair with a fart—unintentional, I wish to believe—impossible to hide as we're not sitting so far from each other. I tell myself it's not his fault because the methane he releases comes from the symbiotic bacteria in his intestine. When you think about it, we are not that individuality we take pleasure in, but rather something like a collective means of transport for bacteria, without which we couldn't survive. Flatulence is the duty we pay, generally easy to cope with.

"...as in this treacherous world,
There is no truth and no lies;
It all depends on
The colour of the glass
Through which you look at it"

Let's not get discouraged then*. It is a splendid evening and I'm surrounded by tropical nature. I believe I've chosen the perfect place to write.

ψ

---

* The verses on the previous page belong to the poem *Las dos linternas* (1846) by Ramón de Campoamor.

Day One

THE SUDDEN SILENCE IMPOSED by the entomologic orchestra that livens up the night wakes me up. The brain detects such changes. I recognise the place. I'm sheltered under the mosquito net hanging from the high ceiling of the bungalow. Yes, I'm in Laos and in a few seconds, the clamour of the monsoon is here all of a sudden with all its clatter. This is the first monsoon downpour I see. It's quite overwhelming but it just gets "bronze" when compared with the torrential drumming I experienced in Monte Alén, in Equatorial Guinea, or in El Darién, in Panama.

Dawn follows night and a bird announces it with a brief, melodious, repeated song; then will come the light of the morning, the heat of midday, some coolness in the afternoon, melancholy at dusk and night again with a shift of musicians. I wish I could explain my ideas in such an orderly direct sequence. The way I see life and mind comes from understanding each of these phenomena and this understanding, in turn comes from my knowledge of Physics, Chemistry and Biology. I'd have never reached these convictions without their concurrence. I say convictions because it's logic that supports the network of ideas that shape coherent thinking. They're neither certainties nor beliefs imported from others.

[Types of knowledge]

Knowledge is a mental representation of reality—not necessarily faithful—that's shaped into ideas and stems from the information we process and keep, so useful in dealing with the issues we face in our human lives. Jorge Wagensberg, an uncommonly brilliant Catalonian physicist, who I read with pleasure and benefit, distinguishes three types of knowledge[2]: revealed, scientific and artistic. In the case of revealed knowledge, ideas come from something (e.g. books), or someone (e.g. authority); they aren't part of our reasoning and are therefore objective. The point is we accept them without testing them. That's why, as Wagensberg remarks, religions remain immutable. Revealed knowledge is never questioned, it's interpreted at most. When we were kids, our parents told us about the Three Magi[3] and many other truths we accepted as dogmas, at least for a period of time. As far as we are concerned, it's objective knowledge because it comes from an external source. We trust the authority, deity or intermediary that delivers it. In short, we believe it to be true.

Scientific knowledge is generated through the scientific method, which is basically testing the hypothesis or postulate that has been put forward. If it confirms it, we accept it as valid. If it doesn't, we reject it as non-true. Therefore, it changes as opposed to revealed knowledge. Some ideas are dropped or replaced by others which are broader or more general and so on. I won't talk about Newtonian mechanics and its replacement by Einstein's theory of relativity because this example might be outside the scope of many readers. It's true that scientists produce scientific knowledge and make science progress, but as explained here, it's also generated by any individual and used every day.

The day the know-it-all friend told us the Three Magi were our parents, we faced a hard hypothesis because the hopeful peacefulness implicit in one of our firm childhood truths was being questioned. Many children must have hidden behind the door to make sure that, indeed, it's parents who walk up to the Christmas tree and nibble at some Christmas cake while they bring in the well-hidden presents, empty the water jar for the camels and drink the glass of cognac left there for Melchior, as we'd been secretly told the poor king was always very cold when away from his warm eastern lands. Thus, putting the case to the test, a belief is dismissed to become scientific knowledge. Presents are not delivered by the Three Magi but by our parents.

I've chosen this example because of the disappointment it implies and because, in a way, it represents what growing up means to us, human beings. With time, many of the truths that were revealed or inculcated in us during our upbringing fall victim of evidence or of our insisting inquisitive minds. We find out that many promises and politically correct truths can't be scientifically supported: they are utopian, fibs, or right out lies. Growing older implies disenchantment, even pain, and we are witness to the crumbling of convenient ideas which made life easier and more balanced. I've often said that happiness carries a lot of ignorance; having no idea is a state of personal comfort. Despite this and when faced with extreme situations (e.g. health) we trust scientific knowledge better, even that which was reached after following the applicable trial and error verifications. Scientific knowledge is objective, or tries to be, as far as possible, and it's intelligible and more compact inasmuch as the idea or formula representing it is simpler than the wide universe it portrays. Wagensberg resorts to Einstein's famous formula: $E=mc^2$, which relates the energy of any material element in repose to the product of its mass and the speed of light squared. Until proven otherwise, this is fulfilled by all the elements in the known cosmos and this knowledge is expressed in just five characters. This is a good example.

Artistic knowledge is essentially subjective, can't be tested and can also be extremely compact. It's born inside us, from our memory, linked to our feelings and on occasion, caused by instinctive pulsion or by exterior messages (music, verses, aromas, etc.) which concisely convey information that's difficult to express otherwise. A melody, a couple of tones, a verse, a perfume can arouse very personal feelings or an understanding which would take pages and pages to describe accurately, if that's at all possible.

What is relevant in this brief synopsis is that people manage daily life using these three kinds of knowledge in different proportions, depending on the individual or the period in life, as none of them encompass everything.

Obviously, technology—an outstanding feature of human beings—feeds essentially from scientific knowledge and has allowed the progress and success of our species. I'll deal with what I understand by success and civilization below, because we must now focus on the difference between what science is and what it is not. I'll borrow a phrase from Horace's Epistles, "*Nullius in verba*" (take nobody's word for it) in the sense that one must not be guided by the words uttered by an authority, whether it is a person or an institution. The phrase has been on the frontispiece of the Royal Society in London since 1663. In a nutshell, trust evidence and proof. And, talking about it, what kind of knowledge do I offer in this book? I can't figure it out myself: more than likely a mixture, but I guess you'll reach your own conclusions at the end. If some of my reasoning convince you, good. If it doesn't, good too. I'm no authority.

Having established this prophylaxis, I come back to the challenge of providing the keys to understanding so that you can follow my personal view on life and mind.

[Organization of the book]

When I read Stephen Hawking's *History of Time*[4] rather than by the space-time theory and the black holes, I was captivated by the way he lays it out, talking about discoveries and the progress made in a sequence that doesn't follow a chronological order but scatters the ideas in a constructive chain that allows the neophyte—and not so neophyte—to put the necessary knowledge together in order to understand the general idea. It's a display of intellectual empathy which astonished me at the time and now overpowers me as, give or take some obvious difference, I'm now facing a similar challenge and I'm aware of my limitations. Shall I put dawn and dusk next to each other to highlight they are mirror processes?

I've started by outlining those ideas I deem essential to carry out my mission successfully, but I get mixed up when it comes to linking them together. I'm even unable to rewind to see what came first in my case, the egg or the hen (or the reptile, which is the right answer). That's why I must once again ask for your indulgence if I tell you things you already know or if you're bewildered by the order I chose to tell them. Unfortunately, chapters can't be read at random in this book. I'll try to lead you by the hand through the maquis I've grown.

Drawing up a script calls for a cold mind, so I put on my loudest swimming trunks and dipped my whitish body in the garden pond stretching before the main bungalow, which doubles as a swimming pool. My feet can reach the bottom and my toes touch the layer of moss that covers it; I'm surrounded by reeds and water papyrus; on the sides and a few metres away, the thick of the forest spreads out and right opposite, I can see the green and hilly valley dotted with ragged clouds. It starts to rain and the water falling has the same temperature as the water I'm floating in and the air around. I'm distracted by so much thermal homogeneity as I'd expected the spur of cold water. I'll try with a beer.

My potential chapters are still on colloidal state, turning and turning around and, to make matters worse, I'm starting to hesitate about the title I had in mind as titles are never innocent: "The meaning of life" Right, but what life? That which surrounds me in the shape of shining leaves, or that I can perceive in the eyes of the frog that follows my movements in its domains without quite making up its mind as to fleeing or remaining contemplative? Isn't it our life the one I wish to talk about, the life of human beings, capable of lucubrating and racking our brains with so many questions? The meaning of our species?... We'll see. The important thing here is that I've already dived into the troubled waters of metaphysics and there's no way out. Let the pen do its job.

[The bulbul and the watch]

Bingo! It's a *Pycnonotus melanicterus* or black-crested bulbul, unmistakeable thanks to that piece on its black head, its white-rimmed eye and yellowish-orange belly in contrast with its darker back in olive shades. It stopped to sing within the range of my binoculars and although my ear is not particularly good, it reminded me of the sickly-sweet singing I'd heard in the morning. So I thought it might play the same role as blackbirds back home, which announce the beginning and the end of the day. The sun has set behind me and I check my wrist watch to note sunset time. Oh, surprise! It stopped at 3 in the morning. I give it some gentle knocks but nothing happens; I wind it, although it is meant to be automatic. No success. Since the kiss of life is no good here, I shake it to revive it. I finally accept it's broken down. This is how far you've come, old friend. It's a Tissot with a black face my father gave to me on my sixteenth birthday. I don't usually wear it but I have it as a keepsake. I often put it on when I travel to countries with modest living conditions where my other watch could be tempting. So I change it out of caution and consideration.

This is the watch I used to wear at La Laguna University to introduce Systems Theory to my Ecology students. It's not a bad idea to start off with the mental hygiene systemic thinking brings. I believe it's interesting and presumably useful in many situations needing to be analysed. So, from the bulbul to the watch and from the watch to the nitty-gritty:

[Systems theory]

A system is something made up of elements which are related to one another rather than to other elements outside it. Expressed like this, it could be a great many things or group of things, both tiny and huge. That's where the fun is; systems don't have a specific size, they're rather like Russian matryoshka dolls *ad infinitum*. I lay my watch on the table. It's a perfectly delimited system. Without having to open it up, we can see several different elements: the strap, the metal body, the crown, the glass face where you can see three hands joined to the same axis on a black plate with twelve lines in a radial lay out, apart from a smaller window that reads 25. With these elements, we can describe this system called watch and, had we analysed it yesterday, we'd have also included its rather simple behaviour: three hands turning at different decreasing pace and the little number we can see in the window moving one unit every twenty-four revolutions of the shorter hand. And by pulling the crown out, we can make the hands turn at will, backwards or forwards. That's all.

If a studious Martian got hold of my watch as the only sample from Earth, he'd be able to describe its system and functioning just as we've done. If he's curious about how it works, he'll end up opening it and looking inside it to find there are many more elements which are equally interrelated and aggregated. He'll identify a coiled metallic band that becomes tense when a loose piece swings (lithium batteries were rare at the time). This tension is dispelled through a balance wheel and a gear train that make the hands move. There's also the number counter. These interrelated elements are, in turn, systems which I'll call subsystems as our reference—the system—is the whole watch.

The one thing our Martian will never find out is why the watch exists; what it is for, what its reason for being is, if it does have one, since it is a gadget. To learn this, he'd need to know about the supersystem it is part of. That is, a human being who divides the time his planet takes to spin in 24 stretches called hours, which make up one day; who wants to know the time at any moment; who has eyes to see (hence the transparent face) and who has a wrist and fastens the watch around it to carry it with him and easily check it. At the same time, he winds it up with his movements.

Jorge Luis Borges has a very good way of putting it: "To see a thing one must understand it. The armchair presupposes the human body, its joints and parts; the scissors, the act of cutting".

This can all seem obvious but it involves strong mental discipline. Once a system is defined at a specific level, the answer to what it is, including its behaviour, is found at that level. If you wish to know how it works, you must go down to the subsystem: its parts; and if you wish to know why the system exists, you can only find the answer by analysing the supersystem it is part of. Asking the question at the wrong level only leads to erroneous interpretations or to frustration. The important issue is where we take aim at.

The hepatic cell, for instance. Its subsystems are the many organelles and inner structures that facilitate its physiology; it is part of a supersystem: the liver. Similarly, we could look at the liver as a system, the hepatic cells as subsystems and the feeding system of our body as the supersystem that justifies its existence. We could thus go up and down scales: from the student to the classroom, to the faculty or the university; from the atom, or deeper inside, to our galaxy or beyond.

Regarding behaviour, that of my watch might seem simple because it is a limited system, with little freedom. Its behaviour is determined by the arrangement of its pieces. This type of system is called machine, but imagine systems which are not so limited, like railway transport, car or pedestrian circulation, each of which has a higher degree of freedom. There are many different types of systems and I'll deal with them in other chapters as, regardless of the nature of their elements, they share many behavioural similarities and by studying one of them, we can learn about the others. The evolution of living beings has things in common with the evolution of language or with the behaviour of our immune system.

One other often tricky aspect in systemic analysis is establishing boundaries or limits for a given system. The cases of the watch or our nervous system leave no room for doubts, but think of the international monetary system, or a dialectal linguistic system. That's a different kettle of fish. In such cases, presumed limits must be established to then check whether the elements within it relate to one another more than to external elements. The input and output analysis—

which must always be done—also helps, and it'll eventually give us an idea of how open or closed our system is (closure levels). Fully closed systems don't exist, except for the Universe, which is closed by definition. If the system has some kind of behaviour—if it's dynamic—it'll always ultimately let out heat. My watch does too although so faintly I don't even notice it.

I'll continue to talk about systems, two in particular: life and mind because they are both amalgamated in us, human beings, and we must split hairs to distinguish their limits. But there are other issues to deal with and I need to rest.

Non lap fan dii.*

ψ

---

* Good night in Laotian

Day Two

IN ZEN NAMKHAN, AND I BELIEVE IN LAOS as a whole, people retire early, one or two hours after sunset. They're active again at daylight. In the West, we continue the day until practically midnight and waste a lot of energy. In fact, the average consumption of oil per day and inhabitant back home, in the Canaries, is about 5 kilograms. With nearly 60% of the population living in rural areas, I don't think they reach a quarter of a kilo here, in Laos. Firewood is still the main source of fuel.

I'm pleased to adapt to the natural pace and I think I'm going to be in mourning for my watch until I leave the country. It's funny, but every time I turn my hand to check the time, I meet my bare wrist and feel a sense of freedom. My dear old Tissot does indeed represent the tyranny of our scheduled modern life.

These ideas come to mind because last night, I sat opposite the lantern in my bungalow to smoke a cigar and wait for sleep to come. The cigar is from La Palma and I brought it with me because I once read Somerset Maugham claiming that cigars had been made to be smoked in the tropics. And he's quite right: humidity gives the tobacco layers a nearly concupiscent texture and the fresh smoke feels like velvet in my mouth. Ah...a pleasure on its way to extinction, I'm afraid.

I was amused by the moths flying around the lamp; big and small, of all kinds of shapes and colours. Do they love light? Suicidal love if it were a candle? No, they're not actually photophilous. They're misfortunate in that now human beings turn lights on while in the past, moths only had the moon or the stars as guiding sources of light for their night flights. If they keep in angle with the moon, they fly in a straight line, because the moon is very far away; but if they're guided by a light bulb, the angle keeps on changing so they try to adjust their course, leading to a spiral movement until they reach its centre point, the light bulb. After some flattering of wings and a few attempts, they end up alighting on the wall or the post, utterly bewildered and captive of their instinctive behaviour. They'll stay there till the break of day unless they end up in the gullet of a *perenquén*[5], many of which also follow lights even though they aren't photophilous. It's food or the expectation of finding it that attracts them.

In such cases, where cause and effect can be mistaken, I always remember an illustrative immortal phrase Ramón Margalef, our great ecologist[6], said: "If a Martian had a quick look at the Earth, he could come to the conclusion that rivers are prone to go through cities".

And this anecdote is a good excuse to get into another matter: time and space scales. I'm not going to get into relativistic physics. First, because it isn't my field; second, because mortals cope with everyday life with Newtonian physics[7]. But I'd like to suggest another exercise in what I call intellectual empathy, even if it means resorting to our Martian friend. Let's play scales.

[On big and small]

A tiny gall wasp—barely half an "i" big— has just taken a walk on the piece of paper I'm using (I recognise its long rear sting). Our scale as mammals is the metre (1-2 m) and the life we can perceive spans from the millimetre of the gall wasp to the elephant I heard trumpeting this morning. Well, the African elephant is bigger, or the 30-metre long blue whale in the ocean, which holds the podium of great known animals, including dinosaurs, which don't exist anymore.

Wow! I've just seen a metallic blue UFO (unidentified flying object) in the distance above the treetops. Its size and the way it moves indicate it could be a beetle and I'm now on the alert. I feel like the war horse on hearing the trumpets, as Darwin would say. Hold it, Antonio, hold it...it's now time to write. It might have been a hummingbird.

Our scale determines and permeates our view of the real world. We think our size is normal and it's actually quite the opposite: we are huge great hulks. Life was bacterial, unicellular, microscopic, for more than two thousand million years. Bulky animals are fairly recent and our size misleads our perception. You regard yourself an individual, a specimen of *Homo sapiens* and you might be surprised to learn that there are mites living in your eye-lashes; that we have over eight million bacteria in our mouth, many others on our skin and above all, in the digestive tube where a very rich bacterial flora lives; and it not only resides "inside" us but also helps us to feed, we couldn't actually survive without them. I enclose "inside" in inverted commas because it's actually an exterior space that goes through our body from the mouth to the anus, and is, in fact, covered with the same tissue our skin is made of to separate and protect us from the external environment. In total, a standard human being weighing about 70 kg has some 38 billion bacteria, which is roughly the same number of cells in our body[8]. They're obviously much smaller, include about 500 different species and altogether weigh less than 250 grams.

Very few people see themselves as a biological collective, a multi-species aggregate belonging to several or different kingdoms. I have no intention of changing the perception you have of yourself, but you should know these facts. Nor are we a pure lineage reached through direct evolution from the first bacteria that ever existed. The same goes for most multicellular beings, that is, the bulky ones. There are functional organelles in our cells which were originally bacteria; the mitochondria, for instance, the chloroplasts of plan cells or the flagella of our spermatozoids. We're polygenic beings, that is, we have genes assembled from other lineages: four in all, according to what Lynn Margulis rightfully coined with the term symbiogenesis[9]. It's as if separate branches in a tree merged to produce a new one. I must say the Spanish Inquisition would have a hard time to issue blood purity[10] certificates.

I say all this because, as the self-conceited species we are, we tend to value ourselves dearer than we actually are. Protagoras said that "man is the measure of all things", and that aphorism is right when referring to humanism but it's not applicable to Science. If we analyse it, the progress made by Science runs parallel to the shrinking of our ego. We started by being special, created in the same image of all-mighty gods; we thought we were the centre of the Universe and that all the heavenly bodies spun around our planet; we then defended free will as something alien to instincts, absolute and defining our greatness. Indeed, our ego as a species has endured a few blows dealt by Darwin, Galilee or Freud. Nonetheless, we do have certain qualities which are not present in other beings or, at least, not as strongly as in us. I'll deal with this later on. For now, a little biological humbleness—our sense of smell is very poor—is convenient to go back to the issue of the scale. Let's push things.

[The moviola of time]

Let's imagine a Martian the size of the moon, whose second of life is equivalent to an hour of ours. If he looked at the

Earth, he wouldn't see us. It's like fast forwarding a film with images of a city. Streets and buildings are motionless while cars and people move faster and faster until they eventually disappear. And if we speeded up the moviola of time even faster, the continents, static masses to our eyes, would be remarkably important and we'd see them breaking off, coming apart or clashing to form great mountain ranges.

You must have taken about three seconds to read this line: the blink of an eye. But, what's happening down there, if we asked a Martian the size of a glucose molecule? Another scale, another time, another world.

Thoughts and the chatter they create in our heads are not ethereal. They're connected to chemical reactions where many neuronal cells take part; not to mention the eyes, which change optical signs into chemical or electrical messages; or memory, which helps us to recover the meaning of what we've read. If we could see it, we'd be overwhelmed by the amount of reactions and cells acting at this level. It all happens in a couple of our seconds, while in the eyes of our molecular Martian, I'd be a stone statue, nearly as boring as continents are to us.

The metabolism of life, what goes on inside cells and their periphery, belongs in a different scale and its time escapes our perception. Our thoughts and the speed they flow at might be the only chance we have to glimpse at their vertiginous dynamism, as seen from our point of view.

In order to understand life, one must be able to move about the different space-time scales involved: molecular (nanoseconds), generational (days–decades), ecological (decades–centuries), and evolutionary (thousands–millions of years).

[The Monsoon]

Hey! The landscape has distracted me again; specifically, the trees around the garden pond and beyond, as far as my eyes or my binoculars can see. Nothing stirs. It's my second day here and it's now I notice how quiet everything is. I ask Loui, a young Dutchman who ended up in Zen Namkhan and is now the manager and factotum; he's in love with the country or perhaps just a victim of his own globetrotter spirit. He says, to my astonishment, this is usual. No wind, no breeze, they only come just before the storm. Well, once again, just the opposite happens on my island where the trade winds never stop, and it's the sudden calm right before nature unleashes all its fury that is quite worrying.

I now realize that at this stage of the text, I haven't confessed yet I'm a biologist; a multi-purpose biologist but relentlessly biophilous. I suppose you'd already guessed it, so please forgive this lapse and my bias. When I travel abroad with my wife, she always asks about the population (6.8 million in Laos) and I, invariably, about the rainfall. Loui checked it for me: 1,600 litres per square metre and year in the province of Luang Prabang. It's not that much if compared with the Luba Crater, on the island of Bioko (Equatorial Guinea), where it rains up to 12,000 l/m^2/year, which explains the luxuriant vegetation it boasts and the common bronze I awarded these Laotian forests.[11]

Speaking of the devil, the wind has suddenly arrived in violent gusts, and Loui has stormed out to fold the umbrellas before it destroys or blows them away. Thunder can be heard in the distance and all of a sudden, my expectations are fulfilled. Trees are writhing like furious giants and the rain is slanting down hard, hiding the mountains in the distance. Several pots with orchids have been blown down and the kettledrums are still playing in the background with the odd lightening. It's the apotheosis I was expecting but somewhat timid. It didn't even last ten minutes and now it's raining meekly while the front moves away.

The water in the garden pond, which looked as if it were boiling and rippled with the wind, looks like a mirror again.

Few drops are falling now and each of them creates a group of circular waves which cross one another, in a random dance. I like watching the waves and I imagine a pond skating bug, those which slide on the surface, trying to work out if the waves are raindrops or an insect and future prey that's fallen in the water. I do hope it's not mistaken by the information it gets from the waves.

The air, which is a fluid like water, but gaseous, is full of waves. We can perceive a very specific range with our senses, the acoustics: thunder, the tapping of drops on leaves, someone talking in the kitchen. But just imagine we had a special electronic machine fitted to our head that could pick up radio, television, satellite waves and all those which being either too long or too short, escape the human ear. What a din!

The air is loaded with information, not only sonorous. Now that dusk is near, in the thicket down there, there must be a hawk moth warming up—beating its wings—to raise its body temperature and be able to fly. Its comb-shaped antennae indicate it's a male. If the downpour didn't take away all the particles in the air, it'd soon trap a very special molecule, a pheromone, sequentially issued by the female. The male will only have to follow the smell trail to find its future partner, which might be one kilometre away. I do hope there aren't many light bulbs on the way.

[Types of information]

We're talking about information—be it chemical, acoustic, visual or sonorous—which most people immediately connect with language and communication. But this is a very elaborate level of information; there are more basic ones which have generally gone unnoticed, even for science. Remember the $E=mc^2$ formula I mentioned yesterday? Mass transforms into energy (heat included) and vice versa. Information also takes part and changes in these processes. But, where does it show in Einstein's formula?

The term "information", polysemic as it is, derives from the word "form" and in Physics it has to do with the way matter and energy are combined. From the moment energy becomes matter and acquires a form—starting by the quark, which is the smallest subatomic particle accepted—it is "informed". Information is therefore a quality of every object or system made up of matter and energy. We call this type of information structural.

You'd be surprised to hear the history a good geologist could trace from the structural information of a stone found at the edge of a road; whether it was formed in the magmatic chamber of a volcano or in the sedimentary bottom of a lake; whether it has been carried by river waters, beaten by sea waves or eroded by the wind. The marks on its surface are also information. If the stone has sharp edges, an animal passing nearby can interpret the risk and avoid it. We can see how this information becomes operational when it's used by an external system. And there's an even more advanced and complex level of operational information: when it's organized into a communication system with the aim to deliver a message from sender to receiver.

The sexual pheromone the female moth released (sender) carries information in its molecular structure and composition, which the male moth (receiver) interprets. Language, our written or recorded messages; genetic codes[12], made up of nucleotides strung together in a long double and twisted molecule: the DNA; the bees dance indicating how much food there is and how far it is, etc. All these are instances of language, there is a semantics that fit both ends, sender and receiver, and therefore messages are operational information. Obviously, communicating through messages involves a previous, long and complex evolutionary process in order to reach such high levels of sophistication.

We human beings are proud of our books and libraries and they're indeed a qualitative landmark of our species to be taken into account. We are now in awe of the USB memory stick, a gadget the size of a fingernail that can store a lot of gigabytes of digital information. But let's change scales and have a look at a human spermatozoid, whose head and all its load of 23 chromosomes is barely 5-micron long (a line of 250 heads can fit in 1 mm). It contains the programme to make a being quite as graceful as the one who produced it, provided that it's matched with the corresponding ovule, of course. I've already pointed out life is small; the spermatozoid you owe so much to is a good reminder. Also, in these 23 chromosomes of the human genome, the record of the history of life is kept, from the very first bacteria. This is a compact historical archive, my apologies to the Library of Alexandria or any other modern one.

I raise my eyes to look again at my boring trees, birds moving in the leaves, an army of insects ready to hit the street as soon as it's dark. How much information do their cells store, their DNA, their gametes in charge of conveying it? When somebody mentions an oak, we always think of the imposing tree, hardly ever of the acorn. I can't recall who I heard saying it, but I liked the metaphor that referred to the seed bank hiding on the ground as the complete dictionary, and the plant community before our very eyes as the words used in conversation. That's what biodiversity[13] is about; a word that's now fashionable and represents the vast information stored in living beings after millions of years of biological evolution, including what we can see and what's hiding.

Information, its dynamics and its evolution in the cosmos has always been just briefly dealt with in the history of Science, especially Physics. It's recently being paid the attention it deserves because, among other things, physicists have realized that we study reality directly but rather the information that reaches us, which is not quite the same. Physicists are very rigorous people and they like being accurate. If we ask a physicist what his dog's name is, he'll say: "I don't know, we call him Pipo".

The fundamental laws of thermodynamics, which underpin our conception of universal phenomena, centre on energy, matter and the heat being captured or released during transformations. Information is unforgivably absent. Margalef outlined a formula that includes it, although he did not bother with it again[14] and I think not enough attention has been paid to it. I do hope a discipline is soon developed — called Infodynamics or whatever — to include information behaviour in physical phenomena[15]. This is the situation and my comments on it end here; otherwise I might stray from my path.

I hope I've explained what information means in physics terms, outside the meaning we give it culturally: data, structured knowledge, communication, power, misfortune... who knows. Joaquín Betrina's moral is well-known: "if you want to be happy, as you say, don't analyse, young man, don't analyse". Can you believe it? Epictetus of Phrygia warned: "only the educated man is free". I want to be free and I'm willing to pay the toll, just like you, I guess.

[Laotian Funeral]

It was already dark, past eight, when Noy, one of the waitresses at Zen Namkhan and Loui's partner, came to invite me to go to a nearby village where the funeral of the cook's father was going to be held. I was happy to accept and shortly afterwards I was riding at the back of a small moped with a determined Laotian young girl shunning holes and puddles down the muddy track from Zen Namkhan. I was relieved when we reached the road as with my 96 kg, likely double the weight of my expert driver, I'd been expecting a great big crash into the ground at any time. Luckily, we didn't fall over.

In a T-shirt, without a helmet and on the back seat, I ended up enjoying the cool night while riding down a dark road across the traffic-free jungle.

The death had occurred a few days earlier and, according to Buddhist tradition, they must cremate the body some days after the family wake and before the full moon. They now invited the villagers and friends to a meal that could go on for the whole night or for even more than a day. They'd built a float of sorts with legs and no wheels, as if it were a palanquin, and inside it they'd put the objects the deceased used during his life. Outside, on a cornice surrounding the cart, which was dressed with a garland of twinkling red lights, they hang small plastic bags with money which they believe could help the deceased in his new life. Old tradition had it all burnt at the end. Nowadays, clothes are distributed among the needy and the money goes to the monastery whose monks conducted the cremation and where the ashes, kept in a jar, may go too if the family so decide. They told me it's quite common to throw them into the river and something else I'd never heard about. Before cremating the body, they put a valuable metallic object (coin, ring, bracelet) on it if the deceased wasn't wearing one. Then, once the ashes have cooled down, several relatives and friends, rummage the ashes until the dead man decides who will find it. The lucky one is thus secured good fortune in his future and in his afterlife, provided that he keeps the piece of metalwork.

Some tents, tables and boards have been set up around the dead man's house, by the palanquin. It's all a bit chaotic. The ground is very uneven; parts of it are tarred, have gravel or clay. I guess the whole village is here, including a few sleepy children. I felt honoured by the invitation, especially as I was the only foreigner in that crowded celebration. As there are three cooks in Zen Namkhan and I didn't want to ask who the widow was, I paid my respects to all three of them with my hand on my chest and a short bow. My way, as I tried to find out what the others did without success. Perhaps to them it makes no sense to give one's condolences.

People cook, eat, chat quietly or play cards. In the background, Laotian music is played but no one dances "to celebrate the deceased" as they do in Equatorial Guinea. Everything is very humble: their gazes, clothes, houses, shacks. There's electricity and one street. Rubbish collection services and the like only exist in Luang Prabang and in the capital, Vientiane. People laugh easily but not raucously. There's a feeling of respect and community.

It's quite clear my height and white hair stand out. They all look at me, some with blank faces and others smiling. It isn't like in Vietnam where people smile as you walk by, as if you were the wind stirring the leaves on the floor. Loui had told me. Laotians are very honest with their feelings; they don't smile out of politeness and if they don't like you, or don't care about you, it shows on their face. Well, I think I aroused curiosity at best.

I've noticed everyone has impeccable black hair and with a salary of 165,000 Kip (some €17) a month, I don't think they dye it. I didn't see alcoholic drinks on the tables either apart from the traditional rice liquor, although this country produces BeerLao, a more than acceptable Lager-like beer. What's quite astonishing and incongruent, it depends which way you look at it, is the many mobile phones there are. Nearly everyone has one.

I had a very pleasant and interesting evening with the Zen Namkhan staff; seven out of the twenty people who work there come from this village of Xiang Lom.

Once I recovered the thickness of my lips, my ears stopped burning and the sweating eased up, we safely zigzagged back on the flimsy moped, while in my head I was going over how many devilishly hot soups I've tried in my journeys in the tropics and hot areas. Another paradox, only in appearance, because hot food helps to fight heat—we cool down when we release the sweat it causes—and it also makes you feel nicely calm after the brutal attack of capsaicin. I believe it even releases endorphins.

ψ

Day Three

I WILL DEAL WITH SYSTEMS AGAIN TODAY, of a very special kind: complex adaptive systems. Although the start has just been boycotted by the UFO of the other day, which turned out to be a very fat beetle with a forked horn on its head. It flew heavily and brazenly across the dining-pagoda where I write, scarcely giving me time to get hold of my swimming trunks and spring to my feet to go after it and try to beat it to the ground. As it came in, it flew out, easily, since there are no walls. I'm lucky there's no one else in the dining-pagoda today and the waitress is sitting with her back to me on the stairs, busy watching a film on her smartphone. What would she think of an apparently respectable gentleman swinging his trunks about? This is no tai-chi, surely.

For many years I've practised tai-chi every morning, as soon as I get up. It's my way of getting ready for the day and keeping my joints supple. I do it at home, in my study and looking at mount Teide in the distance. I've found the perfect place here in the upper pagoda, the library, which is now my office, where no one can see me and I'm surrounded by nature and can enjoy the wood under my bare feet. All the floors are made of a beautiful dark reddish wood and it's a custom here not to wear shoes or sandals in the house.

[Complex adaptive systems]

No more rambling and let's get down to it. Of the many kinds of systems we may find, each of them with their own specific characteristics and behaviours (machines, cybernetics, self-excited, chaotic, etc.), we're interested in those which can learn and perfect themselves. I learnt about complex adaptive systems reading *The Quark and the Jaguar*, by Murray Gell-Mann, Nobel Prize in Physics. He talks about an experiment carried out by students at the Massachusetts Institute of Technology. They built a robot with six jointed legs individually controlled by a computer. Each computer was equipped with a generator of random movements and interconnected to a movement sensor to select those motions which made the whole machine move. The robot would start trembling as if it suffered from Huntington's disease but would eventually move all its legs smoothly at the same time, just like insects do: the fore and hind leg of one side with the middle of the other side and vice versa. Fascinating, isn't it?

The essential issue in a complex adaptive system like this one is that it has a subsystem which tests different options that then undergo a selection process, usually from an external source, to discard the useless and record the useful ones in its memory, so that it can use them in the future. This is how it learns and perfects itself. When the selecting factor changes, the system soon does too, after the necessary tests, and it eventually adapts. Hence the adjective "adaptive" given to this kind of systems, which learn and perfect themselves by optimizing essayed solutions.

Let's have a look at an example from the real world: the mammalian immune system. That is, yours and mine's.

A virus manages to go across the protecting barrier in our body, the skin. It's soon detected as an external invading agent, either through its strange proteins or through the damage it's causing. And it is not recorded. Then, an army of specialised cells—our dear white blood cells, among others—

start to make different chemical substances, one after the other, until they find one that stops the intruder's activities. Bingo! They've made a new antibody which is then recorded and mass produced in order to destroy the invader. The new formula will then be added to the list of recorded defences in our immune memory and it'll be used when the same virus attacks us again. The system learns to defend itself and, in time and with new attacks, will grow wiser and wiser.

Let's move now to language, another complex adaptive system and let's take myself as an example. As I speak no Laotian and the staff here doesn't speak Spanish, we've resorted to English as *lingua franca*. But they have their own plonunciation (issues with the r) and I have mine, I guess. When I ask for "white wine, please", for instance, I notice on the face of the bewildered waitress the message hasn't reached her. I say it again using a different intonation and word order to no avail. I eventually try with a "*plis, guai-uain*", sprawling my mouth, and her face lights up—selective factor—and off she goes to get the elixir of the gods. I catch her drift (linguistic memory) for next time. I even try "*led-uain*" and it works! My language has adapted. I've learnt and I'm successful.

In his book, Gell-Mann also says that complex adaptive systems usually generate other complex adaptive systems. The immune system sprang up from life, which is one par excellence, with natural selection analysing all its tests throughout evolution. But life as a whole, regarded as a system, has many other features that help define what it is. Apart from behaving like a complex adaptive system, it's an open dissipative, self-organizing, mnemonic, replicating, and autopoietic system and tends to expand in space. Don't be put off; I'll duly explain what these strange words mean. But first, I'll deal with what is not life: inert matter.

[Inert Matter]

I've picked up a stone from the garden, slightly bigger than a peach. It's greyish and dotted with whitish spots. I place it on the table and call my metaphorical Martian, used here as an outsider so that we can forget what we know about the subject of our study. He looks at the stone for a long time, both inside and outside; he reaches the crystals it's made up of, its atoms. Everything is in place. Nothing going so he concludes it's a very boring thing, with no behaviour whatsoever, except for the odd molecule on the surface which oxidizes once in a blue moon. This is what we call inert matter, although in certain circumstances, its state can change (ice, water, steam) or suffer chemical alterations due to generally simple lineal reactions, like the oxidation mentioned.

[Living Matter]

I now invite the Martian to have a look at a chrysalid which, like a small lantern, is hanging from a branch in the creeper by the banister. It seems to be just as boring but, ah! he's dumbfounded when he looks inside it and discovers a chemical turmoil. Molecules breaking up, being repaired, growing and getting together to form tissues, etc. This is living matter and its metabolism makes its elements build, destroy and recompose at vertiginous speed. Autopoietic means it's self-supporting, just like you and me. Don't think the atoms that make up your body have always been there; some come in and others go out. 98% are renewed every year. Living beings keep an entity even though all our parts change[16]. That's what living matter does, unlike inert matter.

Besides, living organisms like us grow from a fertilised ovule, or a seed in the case of plants. We self-organize by generating a specific structure which is not created by chance but follows the programme conveyed in our genes, generation after generation. We're loaded with memories (information) from the past and project them to the future. That's what it's meant when we say life is a mnemonic system. The history of an organism, its phylogeny, greatly conditions what it is at present. That's why every specific living organism (living matter operating unit) is more similar to its predecessor than to other beings.

And where does the energy to keep this process going come from?

When I taught Ecology to Tourism students at La Laguna University, who had either studied Arts or Science at school, I'd bring a candle to the classroom on the first day, put it on the table and light it. Having attracted their attention through this eccentricity, I'd ask them:

"Are you looking at the flame? Is it alive?"

Obviously, "alive" is just a manner of speaking, although it does fit in because the flame moves and dances pretty well. The interesting point is that it's an open dissipative system and shares these qualities with other living systems, not other qualities, though, like the capacity to learn or reproduce.

The flame is the result of a chemical reaction called combustion, which consists of combining the oxygen in the air with the carbon contained in wax that is an organic compound just like that of wood cellulose.

Organic compound + oxygen $\Rightarrow$ carbon dioxide + water + light + heat

In physical combustion, the chemical energy contained in the carbon-hydrogen bond of the organic compound is released in the shape of a flame, which issues light and heat. I'd then put a piece of paper over the flame, without actually putting it out, and would remove it quickly to show them a great soot stain, which is recently deposited carbon as it hasn't received enough oxygen to form the carbon dioxide (CO_2), which is a gas and can't be seen.

The structure of the flame, with its dark core, yellowish rim and blue margins, is kept despite the flickering, but if the fuel or the oxygen runs out, the flame is disorganized and it extinguishes.

One of the fundamental laws of physics, the second thermodynamics law, says that everything tends to disorder. As soon as I said it, there was general astonishment in the classroom so I'd resort to a more pedestrian example. At home, there's a small white cabinet with three shelves where I keep the medicines perfectly tidied: painkillers, antiseptics, digestive system, breathing system, etc. If I let some months go by and then re-open the little cabinet, I'll find a real mess in there. So, I must apply my energy now and then if I want to keep that system tidy.

"Oh, yes, it's the same at home". It's a funny way of teaching thermodynamics but they grasped the idea.

The more organized a system is, the further it is from its thermodynamic balance with its parts utterly disordered, as they were before assembling. Any type of order has a thermodynamic debt with the Universe. How come, then, there are beings which are very well-organized, far more than a flame?

Through making a huge effort to go against their natural tendency and keep away from balance. To achieve this, there must be a constant flow of energy to feed the order in the system. If interrupted, the system breaks down: the flame is put out and we'll decompose progressively.

[Mitochondrion and chloroplast]

The energy that maintains us basically comes from the same reaction we saw in the case of the candle; but instead of wax, we use other more palatable fuel: I take out some sugar (glucose and fructose) which I got from the bar and place it by the candle. Now, physiological combustion is not as showy as physical combustion, and we don't burn, even though it does issue heat too (the adjective dissipative means the system dissipates a lot of heat). Biological combustion is called "respiration", in its strict sense, and it occurs within the cells. Instead of light, chemical energy is released and it will in turn be used to build other organic molecules and maintain all the organization of the living being. The water and the CO_2 produced are conveyed to the lungs to be expelled into the air when we ventilate (what we commonly know as breathing). I pick up my glasses, breath on them and, *voilá*, there are the little drops of water just issued. Some students repeat the experiment.

In short, if there's no fuel, the candle is put out; and we, like any other being, die and start to get disorganized.

"What about plants?"

It's exactly the same: they breathe just like animals do but they have a little trick. They don't need fuel to work because they make it themselves. Just change the direction of the arrow in the above formula, take heat to the left side of the equation and you'll have the reaction known as photosynthesis. Plants and algae are able to take water and carbon dioxide and, with the energy from sun light, make glucose (carbohydrate), the fuel they use for their respiration.

This is a great achievement of life but it is no merit of plants but rather of a special kind of bacteria. The organelle or chloroplasts, where photosynthesis takes place, were cyanobacteria which some 1,300 million years ago went on to live within a major nucleus cell, thus inventing, by symbiogenesis, the typical plant cell. In fact, they can't live now outside the cell. The nice thing is that chloroplasts—like mitochondria, which were originally free primitive bacteria—are not built following direct instructions from the DNA in the cell nucleus where they reside; they split, just like any bacteria (they have their own DNA) and when plants prepare their ovules, these carry in their cytoplasm a full kit of mitochondria and chloroplasts, transferring them to future generations via the female plant. And the same goes for animal mitochondria. Males are of no use here. Life is a boxful of surprises, isn't it? My students now have, especially the girls, good reasons to joke for a good while. It doesn't take long before one of them hoists the flag of the mitochondria, as if it were a noble lineage, until some clever classmate, who has followed the lesson, cuts her short:

"Ok, ok, but mitochondria are actually the kitchens in cells"

It's then my turn to cut in and calm things down.

There are other types of bacteria which can use energy from sources other than the sunlight (e.g. sulphur), but I don't even mention it on the first day so they don't get all messed up. I don't talk about viruses not being life either, because it's hard for many of them to accept it.[17] The main thing is for them to learn that plants, algae and bacteria are the first beings to pack chemical energy in organic compounds, that is, they build the biomass other animals depend on to feed and build their own. Sugar made out of cane is my fuel and the wax of the candle comes from bees, which in turn feed on nectar and vegetal pollen. In later classes, I'll explain the food chain (shrimp eats alga, sardine eats shrimp, tuna eats sardine...) and how ecosystems are organized, but we aren't going to deal with that now.

A physical combustion is easily sparked off every time exterior temperature gets high enough to cause a reaction (spontaneous or assisted ignition). However, it isn't the same with life. All the living matter we know comes from the very first time life was "sparked off", some 3,900 million years ago. Since then, it's been transmitted from being to being. It's as if fire had emerged just once in history, and we have had to pass the flame on from torch to torch, like in the Olympics.

Where does the first living matter come from, then? Religious explanations point to a creator who, quite evidently, needed to be alive or something, in order to act. This ends up like a dog chasing its own tail; the same goes for those who think life arrived on earth from outer space. All they do is take the problem and the same question somewhere else.

The living matter we know, with its frantic chemistry, is a system emerging from inert matter, and until proven otherwise, I assume that emergence took place on our planet, leading to all the descendants that make up the present biodiversity, in addition to all that which has lived and extinguished throughout its long history.

I'll continue with emerging phenomena and contingencies tomorrow.

[Oikeiosis]

This place is getting used to me; and I to it. It often happens. You reach a place with certain expectations—or prejudices, perhaps—and it's very unusual to see them all fulfilled. First of all, there's some estrangement because you notice more those things which don't fit in than those which fulfil or even improve your wishes. You then open up slowly, you get used to the noises, the people around you, the light, the pace, the rain... Little by little you recover your centre, you come back to your basic coordinates; you "reset" and start to drench in what is being offered to you; you let yourself go and flow easily with it.

On the lintel of the main entrance to my house in Tenerife, I wrote a word I took from the Stoics: *Oikeiosis*, which has the same root *oikeios* (home, domestic) as economics or ecology and the suffix -*osis* (process), and means something like "become familiar with, appropriate your surroundings". It might be just a coincidence but the name of the bungalow I'm staying in is *Harmony*. From tonight on, Zen Namkhan will be my home.

ψ

Day Four

THIS MORNING I WAS WORRIED for a little while. Although I haven't used it till now, I went to get my mobile phone to check the time; in case it was a good moment to call home (there is a six-hour difference). It wasn't in the usual place. I rummaged about in my backpack, searched the dining-pagoda, my improvised office and my room, including the terrace. I hadn't been anywhere else and as I don't lock the doors and leave things about carelessly, I started to consider the possibility of someone actually falling in the temptation of taking it, even though it's an antediluvian model. I'd noticed everyone here walks about with a smartphone in their hand and Loui had confirmed it's the first thing they buy as soon as they earn a salary. It costs between 150 and 200 dollars and has become the golden calf in this community, followed by the moped. Not to mention cars; roads should be prepared first.

It all turned out to be a false alarm. My old phone was hiding behind a bird guide in a mesh pocket on the front of my backpack. I only use it to talk and it's not connected to the internet as a precaution. Last generation mobile phones deserve deep consideration as I'm afraid we don't gauge carefully enough the gadget we're carrying or the implicit risks in so much *prêt-'a-porter* information.

When I got to the dining-pagoda I was met with an increasingly common situation everywhere. The son of Moon's, the owner, who must be about 5, like my grandson, was curled up in one of the rattan chairs fully absorbed in some game issuing all kinds of alluring little beeps and noises; Lat, the waitress on duty that day, was leaning on the bar looking at some pictures; a bearded young man who looked like a *walkabout*[18] was also typing away on his phone while his soup was getting cold, and the saddest of all: a young couple sitting at the corner table, with a fascinating landscape around them, were hiding their faces behind their phones and not talking. That's how I found them and they went on like that for a while; they didn't even answer my *sabaidii ton sao* (good morning). I'm definitely concerned about it. Smart? Telephones bring people who are far away closer but move away those who are near you. I'll talk about this later on.

An elephant has just trumpeted down there, near the river; I've been told they do that when they're happy. I accept the scolding and get back to work. I'm aware that regarding mobile phones, I'm the weirdo.

[Autocatalysis]

We broke off at the basic chemistry of life which I introduced to explain that an autocatalytic reaction[19]—that is, which is self-animated—is capable of maintaining and projecting in time such an interesting and complex phenomenon as life. Some attribute this responsibility to a "vital factor"—no one has found it as yet—or a "divine blow", idem. I see no magnet or purpose pulling life upward but a pyrotechnic rocket that shoots up once you light it, has all the space to explore and will continue to be propelled as long as there's gunpowder.

[Emerging properties]

Physicists talk about emerging properties when the characteristics of a group of elements can't be explained by those of their individual elements. This usually happens in complex systems.

Just a few steps from by bare feet I can see a line of ants which, like a conveyor belt, are busy carrying several flies and other unrecognizable insects to their nest. Their colony has emerging properties. One individual ant doesn't know how many ants make up its colony, nor how many of them have eaten or not, how many are born every day or how many die in a rainstorm. However, those who have studied it say the colony adapts to its needs so that they send explorers in more or less directions to obtain the necessary food. This is an ability of the colony which individual ants don't have. There are plenty of examples like this everywhere. The important issue here is the concept.

There are whole systems which have emerged from underlying systems, complying with their restrictions and contributing new features. They're more than the sum of their parts and, if the phenomenon is repeated successively, they can end up by forming an aggregate hierarchy.

From this point of view, chemistry can be considered a system emerging from physics. The chemical properties of an atom, its capacity of reaction to others to form molecules, emerge from the most innermost physical qualities, from the mass of the nucleus, from the layers of electrons around it and from many other basic features of matter. The same happens with life, which can be understood as a property emerging from chemistry.

The black-ass ant, the protagonist this morning, complies with physics and with chemistry, but it is far more than pure physics and chemistry; it's living matter and has an (emerging) behaviour of its own: in fact, it is born, grows, dies and disintegrates.

[The origin of life]

In the beginning, some 4,500 million years ago, when the Earth was formed by an aggregation of dust, stones and cosmic ice that was in the area, our planet only had inert matter. These components stratified according to their densities under the rule of gravity and formed the lithosphere, the hydrosphere and the atmosphere.

Not long after that—in cosmic terms—life emerged on Earth. The oldest fossils, proto-bacteria, date from 3,900 million years ago. There are many hypotheses of how living matter came about, starting from a progressive complexity of chemical compounds in a very reactive watery and gaseous medium, in addition to the electrical discharges of lightning. This pre-biotic evolution is thought to have happened in water, which is the only liquid allowing the dissolutions needed for the chemistry of life. It was all set up around the carbon atom, a 4-valence element which is more flexible than its colleague, silicon, especially when it comes to combining to form the chains and rings of carbohydrates, proteins and fats, which are the pieces biomass or living matter are built with. The first major achievement was the emergence of molecules which were able to replicate, the precursors of DNA; the second, the forming of bubbles or vesicles with a proto-membrane, thus separating interior from exterior. This way, the first proto-cell emerged. If there were other attempts to create life with other molecules, they must have failed. From this first pre-biotic phase only one champion was left: carbon.

Life was started like that, modestly, in the water of the time—in the postulate "primordial soup[20]"—after many trials and failures, slowly becoming increasingly more complex until turning into an autocatalytic complex adaptive system which was by then, impossible to stop. These are hypotheses, plausible, but hypotheses all the same. There was no one there to tell the tale and I'm afraid my Martian friend wasn't there either or won't say a word, the bastard.

$$\text{Physics} \Rightarrow \text{Chemistry} \Rightarrow \text{Life}$$

With the emergence of living matter from inert matter biological evolution sets off, a process which is guided by natural selection and contingencies—I'll deal with them below—which introduce complexity and diversity as aeons go by. The result is biodiversity as we know it today, apart from that which was lost in the way, like trilobites or dinosaurs, which our children find so fascinating.

The appearance of the nucleus in a cell as a governing centre, cell aggregation and co-operation to form the first multicellular organisms, symbiogenesis, the invention of sex and programmed death, etc. These are all major landmarks in the history of life[21] we, biologists, know, and can actually wear someone's patience if they give us a free hand. If you're curious about the issue of our origin, there's a book that makes a pleasant read, *La plus belle histoire du monde. Les secrets de nos origins*[22].

What I'd like to do here is to continue working on the concept of life in order to de-anthroposize it as far as possible, as it's widely believed that biological evolution was set up to reach *Homo sapiens*, which is its highest achievement. This is like throwing an arrow and then painting the target around the place it fell. It never fails.

We're wide of the mark. For life to work as a complex adaptive system we have, on the one hand, natural selection which tests every living organism and many of them fail. On the other hand, a mechanism that's constantly generating tests is needed, slightly different organisms, to see which one works best. The most basic, and present from the beginning of life, are the mistakes occurring when the long DNA molecule replicates; some nucleotides may change, add or get lost, thus altering the genetic programme. We call these mutations, they can be fatal and so the organism doesn't survive; they may not be very important and could be left there for future use; or they can be a new trial with adaptive advantages for the individual. Then there's a chance the change may prosper, although in multicellular organisms it will only happen if the mutation has taken place in sexual cells, as these are the only ones which transmit their DNA to descendants.

[Mutations, sex and death]

This type of mutations occurs at random or are caused by chemical substances or heavily charged radiations (e.g. X-rays). Their rate is generally rather low[23], therefore the possible evolution changes are very slow. The other mechanism to generate tests is sex, which is far more efficient; an achievement of life which was selected precisely because of it. Sexual reproduction consists of male and female individuals who produce sex cells with only one of the two sets of their species' DNA contained in normal cells. Let's take the toad I can hear singing on my right; a "love" song, by the way. Even though I don't know the species, its DNA is likely to be aggregated in 22 packages or chromosomes[24], half of which are from its father and the other half from its mother. When its time comes to produce its own spermatozoa, these will carry just eleven chromosomes, but they won't necessarily be the same eleven it got from either of its parents. It'll be a combination of maternal and paternal chromosomes which can generate thousands of combinations, plus the possibility of small mistakes, that is, more mutations that might have happened in the process (losing small pieces of chromosomes, duplications, etc.). Only the combination of 11 chromosomes stemming from the previous 22, amounts to 705,432 possible different spermatozoa. An awful lot. Then, when the spermatozoon fertilizes the ovule, two sets of DNA get together again (11+11=22), so the embryo and future tadpole will get the mutation combinations mummy toad and daddy toad have put in. Also, in Nature, the exorbitant amount of spermatozoids or pollen generally produced per fertilized ovule is no pointless squandering but a means to increase the chance of mutations—new trials—during the copy and segregation processes of DNA.

Humans have 46 chromosomes, apes 48, elephants 56, alfalfa 12 and some butterflies even 380, if I remember well. Just imagine the possible combinations!

There's no doubt sex is a very powerful mechanism to generate variation and new trials. It's virtually impossible for our pair of toads to have two tadpoles genetically identical, unless they're twins. Then selection will favour or punish one or the other and the biological information contained in species will be slowly modified and will incorporate new things now and then. Still, evolution works on a very large time scale—as I said above—despite it having received an unprecedented boost with the appearance of sex. I know that for beings which live under 100 years, it's hard to understand what millions of years involve in evolutionary theory, but there's a little trick. Imagine that the 4,500 million years history of the Earth is represented by the pages of this book. Life would have started on page 29; bacteria and archaea were the only life-forms until page 136; animals started on page 175, mammals around page 210 and our species appears in the last third of the last line of the last page.

The evolutionary achievement of sexual reproduction took place some 1,500 million years ago and from then on, the tree of life has opened and spread with innovations and has created endless kinds of unknown organisms, including some rather big ones. Biodisparity—different life-forms—increases: the first animals (worm-like) appeared 900 million years ago; arthropods 570: plants and fungi, 440. But the compensation for this success was quite special: death.

Bacteria were the first living cells and have no nucleus. Cells with a nucleus and other subsystems with inner organization first appeared 2,200 years later, more than half way through the current history of life*. It's hard to believe, but if we consider their cellular size, it is like comparing the tubby cat that's around here with an elephant; and regarding organization, a small grocery shop with a shopping centre. However, the way bacteria multiply is very simple: they split up in two and distribute its content and the DNA which had been previously duplicated. Did the parental bacteria die? We'd say it didn't. The living matter it was made of lives on in its descendants. So, bacteria don't die in themselves; and if they're well-nourished and an external factor doesn't kill them (e.g. temperature), nothing would prevent bacteria from splitting *ad infinitum* and cover the entire planet. This trend or biotic potential is a feature of self-exciting systems driven by molecular autocatalysis. But the environment and other competitors or predators won't allow it. In other words, there is no room for everyone.

In sexed organisms, two parental individuals engender several offspring, but they're still there. What's the use of trying new individuals if to start off with they're forced to compete with their own parents? It's best to remove the latter once they've fulfilled their reproductive and caring, if applicable, functions. And that's what Evolution has selected. Cells that used to be "everlasting", like our bacteria, now have a programmed death. They can split a specific number of times and that's it. Hence the aging of multicellular organisms which reproduce sexually. We must give way to the young.

From the point of view of information, a system can't evolve much unless it loses part of its memory. The extinction of thousands of species or complete groups of living beings is

---

* In final note n° 20 there is a chronogram with the main milestones in the history of life.

part of this info-dynamic toll. Some "trials" disappear and opportunities come up for new ones. Some people think mammals have prospered thanks to the extinction of dinosaurs in the past, which left no obstacles in the way.

The evolution of life pursues no end, least of all leading to us as the summit of creation. Evolution is blind; once life starts off as a complex adaptive system, it moves forward and makes way sorting and adapting to changes in the environment, whether these are disasters or opportunities. The medium where life unfolds changes due to cosmic influences (solar activity, for instance), to geologic processes operating on Earth (deriving from its gradual cooling, like the drifting of continents) or to random events (a comet colliding...). The tree of life is rather like a bush, without a main trunk, which has been severely pruned throughout its development and it's a result of both natural selection and contingencies. The latter are unforeseeable events which occur and make up living processes and the direction of species: it changes their place, for instance, or destroys them now and then.

[Contingencies]

Contingencies are, somehow, another way to express randomness as far as living beings are concerned. There are two trees of the same species on my left. They have no flowers and their fruit are some chubby berries no bigger than olives, so it'd be too bold to venture to say their botanical family. What's certain is that being individuals of the same species, they share their genome; unless there is a relevant allelic[25] variant, they should be equal. They look equal but they are not. They've both grown in the same environmental conditions: soil, water, humidity and light, but have been shaped by contingencies. A neighbouring tree that falls down and breaks a branch of the tree on the left but not of the one on the right; a caterpillar that eats up the branch shoots that will then stop growing; weevils biting on leaves; fruit that falls at the wrong time... We could count thousands of differences and they're all a result of contingencies. In fact, there are no two identical trees anywhere.

These are small-time contingencies, absolutely unimportant. Let's imagine now the island of Tenerife shortly after emerging from the Atlantic Ocean through volcanic activity, when it scarcely had any vegetation developed from seeds the wind or the birds had brought in from the neighbouring African coast. A floating tree trunk is brought to the coast by the waves. There are two stowaways on it: a pair of lizards in full sexual commotion and a whole island to colonize. They jump out and...oh! A seagull spies them out, dives down and wallop, wallop!, it wolfs them down, unaware of having stopped short a new lineage of lizards on the island. This is a relevant contingency. Some will just call it tough luck.

The truth is life is a system pregnant with contingencies which have led it down one path rather than another, with no other reason but either bad or good luck, depending on how you look at it.

[St. Matthew's Principle]

We've already said life makes progress driven by its autocatalytic chemistry, but there could also be a variational principle governing Evolution. That is, it doesn't only act out of a neutral innocent drive but there's also an asymmetrical factor which bends it to a particular side. It's as if when rolling the dice, there was some kind of trick or rule that gave an advantage to odd over even numbers.

On looking at the evolutionary history of life, although it's the history of winners, there is—my intuition tells me it must be so—something else acting at a larger scale. And this variational principle could very well be information. Margalef hinted at it in some of his writings but didn't go deeply into it, and that's a pity.

When an informed body interacts with a less-informed one, there's no interchange of information to strike a balance, as it'd happen with heat or water in connected containers. It's the more informed body, the more complex one, which can make the most of the information: it'll get more out of it. When two equal computers, one fitted with more programmes than the other one, exchange some data, the former will be able to do more things and take more advantage of it than the latter. This principle, which Margalef called St. Matthew's Principle[26], has many other derivations I'll talk about below. Certainly, living matter stores more and more information throughout Evolution and consequently, its diversity increases. St. Matthew's Principle could also explain why big animals were selected, as they're fitter to store information.

At this stage, I hope your view on life is somewhat different from your earlier notion of it; I hope you understand it as living matter, an extraordinarily interesting phenomenon we're part of but not the main character[27]. This was my intention and if I haven't been successful I'd rather not know. This way I won't get depressed and I can go on writing. Advantages of you being there and I being here.

ψ

Day Five

MY INTERNAL CLOCK HAS ADAPTED to the local light cycle, so I now wake up at five thirty, just before dawn. At this time, birds are most hungry and are very active. In my journeys, I usually carry a pair of binoculars and a bird guide of the country I'm visiting. There's always a chance to enjoy myself watching birds, which are a remarkable source of beauty. The bad thing about very leafy forests—I'm surrounded by one—is that birds move about the leaves and you can hear them and follow their hops but can't see them from the ground. At best, you glimpse at their outline or distinguish the colour of their belly, apart from running the risk of ending up with a stiff neck.

In the humid forests of Costa Rica I found out that when you crouch due to physiological needs and spend a while concentrating on your purpose, everything around you starts to move and you can see more birds, rodents or monkeys than you ever dreamt of. It's as if on seeing you so busy in such a biological occupation, they definitely accepted you as a harmless mammal and a forest comrade, at least while the operation is in progress.

[The Namkhan River]

I went down to the bank of the Namkhan River—a few minutes from my bungalow—with the hope of seeing the elephants I hear going by every morning. The river is some fifty metres wide and there's a landscape of staggered mountains on the other side. Low mists doze off in the small basin cladding it with that magic touch that's so characteristic of sunrise in these intermediate jungles. Nothing stirs except for the gentle flowing of the water. I remember noticing yesterday the smoke coming out of a fire some labourers had lit to heat up their lunch. It went up in a vertical column like the wake of a rocket. There must be a secret in this valley to explain the lack of breeze. I don't know its name but for the purposes of this story I'll call it the Calm Air Valley.

Slightly removed from the bank, I bumped into a Japanese-like pagoda hidden in the forest. It has sliding screens and a wooden platform that leads to a small pond covered with water lilies; it's all surrounded by vegetation with the river at the front and the mountains behind. It suits me a treat! Thrilled with the place, I take the chance to do my tai-chi exercises, and by the by, tune my hinges; or line up my chakras, as some would say.

A bit later, some long canoes go past. Very low and narrow, they're driven by a pole or by a small outboard engine with a very long axis crowned with a propeller that doubles as helm. On one of them, three circumspect locals float away, the one in the middle is smoking a cigarette; on another one, a tiny old lady pulls it to the shore and downloads a huge grass bundle, is it for elephants? The oddest one was a lonely fisherman who, after pushing the pole, would raise it and dropped it flat on the water again with a loud crack. It might be to frighten fish and lead them to some strategically positioned net. Tough work, mate!

At least I managed to identify a couple of bird species and saw a striped squirrel that shot out of the way as soon as it noticed me. I can't reproach it as, judging by the shots I hear at night—sometimes during the day—every big enough animal here is fit for the pot.

I spend two hours letting myself sink in the tempo of nature and rural life in perfect synchrony. I'm waiting for the elephants but they don't turn up. Eventually, like in previous mornings, I can hear the *mahout* or elephant drivers, shouting instructions, but the voices come from somewhere down the river or from the other side, behind the thick wilderness that conceals them. The river is swollen by the monsoon and the paths bordering it are flooded. I won't have the parade I'd hoped to see from my comfortable viewpoint and decide to go back to have a refreshing shower and enjoy a good breakfast. The elephants have not trumpeted this time.

[The biosphere]

Yesterday I finished with the emergence of living matter that eventually took up all the hydrosphere (where it originated), a strip of the lithosphere and a good portion of the atmosphere. This mantle or wrapping of the planet where life takes place is known as biosphere, although the organisms themselves don't take up much space. I read somewhere that if we put all the beings currently in existence in a great blender, we would get a green dough that would wrap the Earth with a four-centimetre-thick layer. Not much, is it? But it has stirred things up, indeed!

Having looked at what living matter is and, in broad strokes, how it works, we should now approach the reason of life as a natural phenomenon. What is its purpose? Remember we must look at the supersystem it is part of, which is none other than the cosmos.

The first question is to find out whether there is life in the cosmos apart from on Earth. We don't know. The second question: are we facing a unique phenomenon, exclusive to our planet, or is life just a stage in the evolution of matter?

[Theory of chaos]

To support the first hypothesis there is a possible argument in that life is the offspring of chaos. I don't mean chaos in the sense of complete mess, total disorder or *kilombo*, as an Argentinian would say, but the concept derived from the Theory of Chaos[28] in Physics, which deals with apparently erratic, unforeseeable behaviours which are present in some supposedly determinist dynamic systems and depend greatly on initial conditions. Small variations in these initial conditions can imply great differences in future behaviours, thus making long term predictions impossible. This usually happens when there are many variables involved. The unforeseeable and unrepeatable traffic jams are a good example; the eddies I saw this morning on the surface of the river; or the water jet that regularly falls on the garden pond right opposite me, which suddenly falls into chaotic turmoil, noticeable because the sound changes...

The thing is Earth is in a window of habitability within our solar system. It has mass enough to retain by gravity the water and the gases that make up the hydrosphere and the atmosphere, which would be lost in space if its mass were lighter. Its distance from the sun is neither excessive nor insufficient for the temperature on Earth to have liquid water which, as mentioned earlier, is the essential solvent for the chemistry of life as we know it.

On the other hand, the Earth is not right on the centre of the window of habitability but is rather close to its outer borders. Boundaries are very interesting because they're close to a change of situation and these areas are neither too stiff to prevent the appearance of changes nor too loose to make the elements independent, causing the system to fall apart. It's an area that fosters innovation, emerging phenomena and chaotic behaviour. That's why some believe that, under these circumstances of "liberalism", life might have emerged marked by the specific and probably unrepeatable conditions of the moment. That is, if another life randomly emerged, it'd probably take another path, utterly different from the one we know.

It's an option to be taken into account although, personally, I'm inclined to believe that life is a stage in the evolution of matter in general and the evolution of information in particular.

[Rising complexity]

If you stop to analyse the history of what exists, you'll notice that from the start, matter has not stopped becoming more complex and gaining structural information. From the first concentration of energy to form the first material particle—the quark—following the presumed Big Bang 13,800 million years ago, everything has turned increasingly complex. Quarks are grouped together to produce different types of particles; these to form atoms which in turn form molecules. Obviously, the first atoms were simple, like hydrogen and helium, and then in the nuclear fusions that took place in the stars, heavier atoms were forged (iron, lead, etc.) which would then be released when they exploded and roamed the space as star dust, until gravity grouped them together again to shape heavenly bodies like our planet, which is second generation (9,300 million years after the Big Bang). That's why astrophysicists say, and rightly so, we are star dust, to give a romantic touch to a profession many people mistakenly deem arid and removed.

The progressive increase of matter complexity is coupled with synchronic information. With the emergence of life, this increase speeds up. From a few thousand minerals in inert matter we went on to millions of organic compounds generated by life. Remember our immune system—in fact, one of the most creative things known to us—capable of making up an endless number of complex molecules which were previously non-existent. A blooming of biodiversity, exploring new forms of life and ways of doing things; storing the information of their achievements and even of their non-achievements in some cases (they're exposed to potential future use). In this context, life can be assumed to be a step further in the complexity of matter and the acceleration of the way in which information flows (its dynamics increase). In the global scenario, this could imply the Universe is gradually uploading information in exchange for expanding and cooling down after its hyper-hot start with the Big Bang. In fact, heat destroys information and in a close system like the Universe, it can't dissipate.

So accepting this second hypothesis, I see no reason why the same or very similar thing as on Earth may not be occurring in other parts of the Universe, give or take some details. And it pleases me to think, whenever I'm watching the starred night, that there's a myriad of planets with life on them, as many or more than the lights I'm looking at. Not necessarily biped and conceited life like ours, but systems that fulfil the requirements you know (adaptive, autopoietic, mnemonic, etc.).

[The meaning of life]

If this is the case, the meaning of life derives from its function in the cosmos and is as simple as to keep on operating and carrying out the exploration of new paths for matter and information—life is a state of matter—, however restricted it might be to the habitability limitations of the planet in question.

The meaning of every living being, including us mammals, is to practise life and pass it on, pass the torch on to our descendants so that the flame is not put out. Procreation, which in our case also involves upbringing.

Simple and crystal clear.

Now, once the meaning of life in general and that of our biological life as a species has been solved, there's another pending issue, as we're more than living matter; we are also thinking matter. My story will go on down this path, but not today.

ψ

It is Saturday, I'm the only guest at Zen Namkhan and I'm enjoying absolute peace and quiet. I keep on writing in the pagoda above the dining-room which, like a watchtower, overlooks the sunny landscape of Calm Air Valley. It all looks like a still photography. Well, some clouds in the distance slide slightly and the mirror of the water in the garden pond ripples now and then, when touched by dragon-flies of a delicate lemon green colour, or when an unlucky flying insect falls on it. Some butterflies enter the scene with their erratic flight, to confuse predators. Some movement, at long last. I've counted some nine species and they're all beautiful, like the flowers they go to. The most spectacular one looks very much like our monarch butterfly but with bluish instead of orange hues, and much broader wing veins. It shows off openly before this eager entomologist, who hasn't caught a beetle worth admiring yet. I haven't confessed it but coleopterans are my favourite; I just love them, my apologies to ants, which I greatly admire too, although I don't study them.

[Hamza, the Moroccan]

Two Laotian girls are downstairs; the others must be on their day off or it's not their shift. They have one day off a week, when suitable. I can also hear the hammering of the carpenters who are making a new bungalow; ten hours a day, Loui said. Holidays, minimum wages and other basic labour rights are still to be developed in this country, or at least in this area. Any trade unionist from the West would be appalled at the rate of staff replacement in businesses like Zen Namkhan.

I didn't write one single line yesterday afternoon. The bearded young man who was fiddling with his mobile phone at breakfast returned from an excursion to the waterfalls nearby. He'd joined in with the hope of seeing a virgin forest, but they took him to a rubber tree plantation. His name is Hamza and he's Moroccan, from Casablanca—we're virtually neighbours—medium height, heavy-built and quite dark. He has long, curly black hair, and a well-kept beard wide at the sides, the same cut as his king, Mohammed VI. He has very dark Mediterranean eyes and a permanent smile, as he rarely closes his thick lips, very handy for his inquisitive curiosity. He must be about 25 and finished Maths in Belgium, where he's currently living. Now he's travelling through some south-eastern Asian countries as part of his personal education.

We start our conversation with some small talk; his visit to the Canaries, my visits to the Moroccan Anti-Atlas and the Sahara; I told him about the Think Global School[29]—where my son worked as Maths teacher—and many other things; we rambled on, he told me about his experiences and aspirations and I told him about mine, as it's usual in meetings like this. And then we got to my stay in Laos and this book.

Insatiable blessed curiosity this young man had! Late into the night I had to explain to him the essence of what I'd written so far and the issues I still had to deal with; I had to give him many examples and tried to answer his doubts, criticism and the paradoxes he perceived. It was testing ground and I'm honestly grateful for the time we shared as it helped me to detect points which needed more attention and to convince myself that this literary adventure, if I may call it so, is worthwhile. He did recharge my batteries, no doubt.

The idea of the psychosphere (more of it later) he thought very original, like everything that changes your point of view so that you see the same under a different light. This was the most recurring topic. However, when he came to say good-bye early this morning, after agreeing that it'd been a most stimulating evening, he confessed that it was the concept of contingency that had really made him think. That's odd. He also told me he was up very early to write the first pages of a travel notebook, something he'd never done before, had not even thought about it.

"You've been a contingency to me!" was his brief farewell.
"That's good, Hamza".

ψ

Day Six

IT WAS A VERY FOGGY DAWN. I'm not surprised as it rained heavily last night and fog here comes from drenched soils; it's not like the *nublina* on my Islands, low clouds touching the mountain slopes. A few metres away, trees are blurred, like outlines of dauntless and watchful spectators, like me, of the birds' concert. This is a shot of chlorophyll directly into the blood and it feels wonderful as a starter for the day: the main course today is the mind.

It's no easy topic, as not everyone can tell the difference between intelligence, conscience, and mind; I know there's still controversy among biologists, psychologists, psychiatrists, neurologists and theologists. So, I'll tell you my view, with a little help from Wagensberg[30], who has recently dealt with the issue of intelligence with his characteristic systematic lucidity.

[Rising communication]

The nervous system appears and develops in the evolution of animals as a need to manage the exterior information perceived by the senses, when moving in search for food, in order to avoid unfavourable or dangerous situations, or when searching for a partner. Algae, fungi and plants can respond to external stimuli like light, temperature and humidity; they're even capable of communication but haven't developed a nervous system with specialised cells (neurons) because they don't need to travel.

The forest around me is not a collection of independent trees competing against each other, exposed and defenceless to the changes in their environment. There's a fungi network under the ground and their kilometres of mycelium connect trees, in a similar way to the internet, and allow them to co-operate by exchanging water, carbon, nitrogen and other nutrients or more complex substances, like some hormones that stimulate the defence mechanism[31].

There is a real communication network with no differentiated nervous system. By the way, fungi are very original organisms, neither plants nor animals, which develop in the terrestrial medium, feed on organic matter—usually dead—, helping to decompose it and form soil. We usually only notice the mushrooms, which are their sexual organs. It's as if I was buried in the ground with my willy sticking out. The main body of fungi spreads underground in the form of mycelium, a branched network made of very thin threads— the hyphae—through which they feed, while at the same time serving as a horizontal conveyor of minerals and other staple substances; the motorways of the soil. There are many fungi which co-operate symbiotically (mutual help) with plants, forming mycorrhiza (literally fungi-root); so vegetables— especially the biggest ones, trees—are not as independent as we think.

There's room for everything in animals and Wagensberg was right to establish levels or degrees of intelligence, which helps to understand.

[Levels of intelligence]

Level zero of intelligence, that is, its nonexistence, is found in a stone, in inert matter. The information that reaches stones bounces, it floods or goes through it but nothing happens; there's no capacity to process it. Inert matter is exposed to the events in the environment and it can't anticipate anything: totally defenceless.

The ant that now and then strolls round my table which I now find quite nice, can respond to specific stimuli although it does so automatically, following "fixed patterns of behaviour", because these are recorded in its genes (instinct). I remember an amusing experiment I read in a book by Bert Hölldobler and Edward O. Wilson entitled *Journey to the Ants*[32]. When an ant dies, the decomposition of its body releases oleic acid two days later. If the death occurs inside the ants' nest, the other ants get rid of the corpse, taking it to the chamber used as rubbish pit. Wilson had the idea of covering some living ants with oleic acid. A little later, their fellow ants grabbed them and despite their frenzied kicking, dragged and threw them to the rubbish pit. Every time they tried to come out of it, they were returned to the rubbish pit. You're dead and that's it!

This is level 1 of intelligence and Wagensberg gives the example of a squid which is offered a crab in a jar. It'll try to reach it head on, banging against the glass, and it'll never occur to it to do something else, like opening the jar.

However, the octopus, which is level 2 of intelligence, at first tries to reach it just like the squid but when it realizes this is no good, it tries other methods until it manages to open the top. And it doesn't forget it. The next time a crab in a jar is offered to it, it goes directly to the top, and opens it to eat it. The new plan is now part of its personal culture, and I say culture because this piece of knowledge is recorded in its memory but not in its genes. Its descendants won't inherit it through its sperm; or its ovules if it's a lady octopus.

Near me, Meggy is lying down on a cushion in a comfortable armchair, as she does every day ever since I arrived. I can't imagine what cats see in me, being such independent and disdainful animals, but they're always around me, despite my not giving them any food. I remember Poisson, a yellowish cat with a white tummy and mottled hind legs that adopted me while I was at Gump Field Station of the University of California on the island of Moorea. But Poisson (fish in French) was cheekier and loved to climb on my desk and tried to play with the tip of my pen while I was writing. Meggy is very lazy and when I go up to her and scratch her belly, she doesn't even open her eyes; she just stretches her forelegs a little and starts to purr. Well, that's a behaviour characteristic of intelligence level 3.

Cats' and dogs' instincts are not triggered automatically; they can control them based on the available information, either present or memorised. Meggy, for instance, won't leave a smelly souvenir on the cushion. On a physiological need, she'll control her sphincters and wait till later, when she's in a place where she won't be told off. That's not the case of geckos, whose bombardment of droppings from the ceiling make a real mess; or the souvenirs a buffalo would leave if allowed to walk about the garden.

Animal Intelligence Level 4, which involves the use of tools, is not very common and I missed the chance to see it during a project I directed in the Galapagos. I knew of the existence of one of the so-called Darwin's finches, the woodpecker finch, whose tongue is very short so it picks up a little branch or a cactus thorn with its beak to help remove larvae from the holes in wood. I spent quite a long time on the Island of Santa Cruz but was not lucky enough to see it in action.

There's an illustrative documentary by the National Geographic Society[33] that shows how chimpanzees experiment with and use tools to catch insects, open nuts, drink water or just play. For example: a chimpanzee introduces a small stick in a safari ant's nest, which will furiously attack it; then the stick is taken out packed with nutritious ants grabbing it with their jaws and the chimpanzee slides it across the lips to eat them all, without them having time to bite it (our relative seems to like hot meals too). Other chimpanzees in the group see the trick and, great imitators as they are, will try the same technique until they're successful. This is an instance of cultural transmission among individuals; an instance of simple social culture.

What chimpanzees won't do is keep the stick and use it next time they see an ant's nest. To do this, one needs to be capable of abstraction, understand oneself as an individual and project one's image in time. One needs to be fully aware of one's own existence, situation and the consequences of one's acts in order to plan the future.

These abilities are found at intelligence level 5, where human beings are. There are signs of awareness in other animals, as seen in the studies of self-recognition with mirrors, recorded in great apes, dolphins, elephants and magpies[34]. But it's incipient awareness, not enough in itself to support intelligence level 5, or we should need to add a new level for the capacities of human beings. Consciousness goes far beyond proprioception—sense of the physical limits of one's own body—present in nearly all animals and probably a gradual process developed from the latter. The problem with this issue of consciousness is that the feelings of people are involved, including many scientists who attribute consciousness and feelings to animals, an area of troubled impassioned waters, which is beside the point here.

Insects have most of their brain spread across the segments in their bodies. Octopuses' nervous system and eyes have evolved separately and independently from mammals. These cephalopods have a central brain and a minor brain at the base of each of their tentacles, eight in all, with capacity to make decisions, as if it were a federal regime. Humans have an enormous neocortex, compared with the size of the body, apart from the so called "avian" (behaviour management) and the "reptilian brain" (basal working of the body) plus major neuron groups in the heart and the stomach which have certain autonomy.

[Thinking matter]

The highest degree of conscience in human beings is found in a very powerful brain (with neocortex) capable of managing a complex, highly developed language which permits abstract thinking leading to such different results as art, science, advanced technology or morality. Important anatomical modifications have been necessary in the larynx and the breathing system[35] to allow speech and the development of language—words—required for abstract thinking; these achievements come through an evolution that has run parallel—mutual support (co-evolution)—to the neocortex.

Bear in mind what I explained about information and the size of things. It's therefore no surprise that in a brain the size of ours, with cutting-edge intelligence, the capacity of vocalized language and an incipient conscience, a phenomenon like the mind should emerge. By analogy with inert matter and living matter, I like calling it thinking matter. No conscious matter but thinking matter because that's basically what it does; it manages and generates information (ideas) as it had never happened before and, in fact, it does so unconsciously too, and when we're asleep. The mind is an emerging system of biology, just as the latter was of chemistry, which in turn emerged from physics. It fulfils all the restrictions involved in these subsystems. But it's something else, definitely something more than just conscience and occasional feelings.

$$\text{Physics} \Rightarrow \text{Chemistry} \Rightarrow \text{Life} \Rightarrow \text{Mind}$$

Thinking matter is recognizable through its behaviour and its effect on the environment, just as it happened with the emergence of life. Thoughts, both words and images, are recorded chemically—some may be reset when remembered—even when we sleep.

All this involves a tricky chemical hustle and bustle our Martian friend would not notice at first. Let's place our Martian very far away from the Earth; in Jupiter, for instance, 25 million years ago, before hominids existed. Not long after starting his study of our planet, he'd conclude something odd was going on. The gas layer around it has an abnormal proportion of nitrogen and especially of oxygen (22%), which has no logical explanation in chemistry. Oxygen is a tremendously reactive element and should be combined with other elements to make oxides, carbonates or just carbon dioxide (CO_2), which is, in fact, what there was in the beginning.[36] But that's not the case. There's a trick: someone or something keeps an absurd chemical imbalance. If he comes closer his suspicion will be confirmed: living matter is removing carbon from the atmosphere and pumping oxygen ever since it achieved photosynthesis. Ok, this planet has a biosphere and that explains everything, even its blue colour. The Earth's original sky was white; oxygen has turned it blue and it reflects on the oceans.

If we change the date of the visit to the present day, our Martian friend would be surprised way before even entering our solar system, beyond our neighbouring stars (Alpha Centauri, etc.). All of a sudden, the receptors on his interstellar airship would start to pick up music, spoken messages and images. It wouldn't take long before he realized all that din in space came from the Earth. Radio and TV broadcasts; highly structured information…

"Well, I've bumped into a planet with psychosphere", he'd correctly conclude.

[The psychosphere]

I might have coined the term psychosphere[37]. There was another one introduced by Le Roy and Theilhard de Chardin, noosphere, which means the sphere of intelligence (nous), but Vladimir I. Vernadsky, who made it popular, gave it a teleological and possibly naive nuance: the advent of a planet ruled by human intelligence. I wasn't convinced by the concept of anthroposphere either (*anthropos*, human being), used by Josep Peñuelas[38] to describe a biosphere exploited by humans. What is qualitative and relevant, really new, is thinking matter, whether it's one species that holds it or any other that reaches the capacity of thinking with global consequences. Although the original Greek term, *psyqué*, means human soul, I apply it here to refer to all the processes and phenomena that make up the [human] mind as a unity, the same root as in the word psychology.

There were other species within our genus (*Homo neanderthalensis* and *Homo antecessor*), now extinct, who left enough funerary and artistic signs so as to attribute them a mind. Therefore, the start of an incipient psychosphere—also gradual, as with the biosphere—can be dated back to some 75,000 years ago (first artistic representations). There were other species which used fire and built tools much earlier (*Homo habilis, Homo ergaster, Homo erectus*, etc.) as appropriate to a level 4 intelligence. The number of hominids recognised as a species different from *Homo*[39] is a soap opera that may have too many scriptwriters. The merit of *Homo sapiens* certainly is survival and at present, being the only *lucifer* (light bearer) of understanding.

We don't know if other animal species with a complex social life are far or very far from reaching the brain "ripeness" to allow the emergence of a mind. Social insects—termites, bees, ants—seem to be too small, unless they manage to interconnect their hives and go up a few steps in animal integration: octopuses, being so clever, only live for two years, which I see as a poor result for such advanced neuronal organization; birds would need to develop a neocortex and mammals—great apes and dolphins—would be the best candidates for the emergence of new minds. But I guess they still have a long way to go in the evolution of neuronal complexity and morphological modifications before they can develop at least a symbolic language based on acoustic signifiers that enables abstract thinking.

[Life - mind differences]

There are many differences between life and mind, as it happens when a system emerges from another one, but this is particularly relevant for my discourse. Life originated once[40]—if there were other attempts, they perished—and since then it has been transmitted from living being to living being, without extinguishing and its information channel goes through a bottleneck: the genes. In contrast, the mind emerges in every human brain once it reaches the appropriate ripeness. What does this ripeness consist of and how does it happen? No idea. Finding out is probably more complex than experimenting with the primordial soup and getting a few organic compounds in a laboratory. I guess researchers on the so-called artificial intelligence are trying to make a machine that not only mimics our way of thinking but that can also think from self-awareness. I don't know either at what stage in the development of a foetus or during our tender childhood consciousness merges with cognitive capacities. What I do deduce and assume is that mind is not conveyed in spermatozoids and ovules as it happens with life. Our gametes are no thinking matter. What's passed on in our genes and cytoplasm are the achievements acquired through evolution, which allow the formation of the anatomic-neurological-physiological setting where the mind will emerge: one per every human brain.

[Souls and gods]

I'm aware that these ideas may be awkward to some readers, especially to those who believe we have an immortal soul that doesn't succumb once the living matter stops working; a soul that comes or is given to us from an outer source, either by a divine blow or because we are attractors of a major whole and will actually go back to it. In my reasoning, I don't need this kind of explanations, nor do I intend for believers to stop believing. Every belief about the human soul is compatible with my explanations. But I'm an atheist, and I stress atheist because a believer, I am: I believe in the inexistence of all-powerful gods in charge of the whole show who occasionally reward or punish. And I believe it because I can't prove it.

[Ideas and creativity]

Returning to thinking matter, its merit lies on how fast it manages information and its capacity to generate new information (creativity): abstract, intuitive ideas by comparing similar things, or by combining or extending pre-exiting ideas, whether they come from nature (bio-emulation), from our mammal instincts, or from events recorded in our memory. While the immune system of mammals is extremely creative in terms of chemical substances, the mind is creative in terms of operational information; that is, ideas capable of solving problems in life, reason enough to have been favoured by natural selection.

Don't think ideas are immaterial; they all have material support like every type of information. We'd need to give a huge scale leap to rub shoulders with molecules and the molecule aggregates that compose them, which react during the act of thinking, in the inter-neuronal spaces, in the glia and in the neurons themselves. When I think about the brain it's just like when I look at the sky on a dark starry night, I have the same feeling of infinity[41]. I get lost in this universe although I know that thanks to technological progress, neuroscience is starting to uncover the deep mysteries of the brain. Not to mention hypotheses like those on morphic fields, which connect information and modulate it even beyond time and space scales. The idea of morphic resonance[42] is attractive because it would explain many mysteries which persist in current science, from causative formation in the ontogeny of a living being to quantum entanglement. But it'd be cruel of me to lead you down the path of subsystems of subsystems of subsystems. Let's stop at watch level, it tells the time.

[Cultural evolution]

Life transmits its achievements basically through genes[43], but it has achieved great advances by combining different beings, like in mythological chimeras. I mentioned on Day Four that the cell with a nucleus was originated through fusion (symbiogenesis); that plants have five genomes from different lineages and animals have four; that part of these genomes is carried in the ovule cytoplasm (e.g. chloroplasts and mitochondria); in short, in life, information is handed down from generation to generation.

With the mind, information passes on from one individual to another or is recorded in external memories—stone engravings, books, computers, etc.—so that they can be retrieved in other places and at other times and, potentially, by anyone. This is new and that's why cultural evolution is far faster and has more creative potential than biological evolution.

[Will and finalism]

Determinism or finalism is a new phenomenon which appears in the (known) Universe after the emergence of the mind. Up till then, the achievements found in Nature are the result of evolutionary processes and contingencies, with no end or purpose whatsoever. It's humans who create ideas which project into the future; plans appear to achieve a goal followed by the determination to reach it. There are many new artificial things now,—like the fountain pen I'm writing with—achieved through will and the action of the mind and not as a result of a blind process. I use the term artificial to refer to all that related to human beings and their technology, as opposed to natural things, which are there without us interfering. You may point out—as Hamza, my young Moroccan sparring partner did—that humans are just as natural as any other animal, for they came from Nature; true, they did. But then we'd need a non-existent adjective to refer to non-anthropic natural things, those which are naturally natural, or we'd have to explain it every time the topic comes up; that'd be very convoluted for my piece of writing. It's a matter of semantics and I hope you accept the antagonism natural versus artificial / anthropic, at least for practical purposes.

When I spoke about determinism as a cosmic novelty, Hamza raised a brow; and when I added we needed a new Ecology to deal with it, he frowned. So, I looked for an example, which is normally very useful.

"See that bungalow they're building below the garden pond? There are some hundreds of bricks there, carved stone that comes from some nearby quarry, wood that comes from some felled trees which were then sent to the sawmill and transported; there are wires, electrical light bulbs made in China, perhaps; there are pipes and sanitary ware, which were in turn assembled with components coming from many different places. And it's all being organized into a functional unity thanks to the energy of a few labourers who distribute materials, mix cement, cut and screw pieces together following the plans made by an architect with the design of a bungalow fit for guests; and all because Moon, the owner of the place, wanted it and arranged it. Do you think, dear Hamza, that such an aggregation of materials would form in Nature by itself, however capable of organization ecosystems are? The bungalow you can see results from human will introduced by the mind, which searches for whatever is needed to achieve its goal. Sheer finalism" – I concluded, and I think he was sort of convinced.

[Information ecology]

Ecosystems are self-organized but they don't follow a pre-established plan. They're the result of the interaction of present species or those that may arrive, each of which develop their own programme and settle down—as well as they can or are allowed to—among the others. The outcome is a very structured functional system, reached through different phases where some species find their slot and facilitate others', following one another in time; the information in the ecosystem increases gradually and the energy which makes everything work is used more and more efficiently. So, it's an orchestra that sounds wonderfully with neither score nor conductor. It's the musicians themselves that achieve harmony; no one goes out to hire a predator to make the symphony sound better. That's the biospheric ecology we know and have studied at universities (in the good ones, that is).

Imagine I had your phone number and called to ask you to go to the garden and hose the plants. You do have a garden and follow my instructions. I've employed a few milliamperes to convey a message from this part of the world to the ecosystem wherever your house is, and plants have just received additional water that doesn't come from the rain. Such a phenomenon is not contemplated in classical Ecology. Again, it's information that escapes us. Now that our mind is working nonstop and at a major scale, we need more than ever an Ecology that considers the psychosphere as its new subject of study. We need an Information Ecology. How many things are happening on our planet now as a result of the highly structured information on the Internet?

In the last few months, the relationship between Washington DC and Pyongyang, in the People's Republic of Korea—on this area of the planet—has not been particularly friendly and, both having quirky rulers, the issue could end up taking its toll. Imagine it isn't me making such an innocent call but it's Mr. Trump or Mr. Kim Jong-un calling another number and giving the order to...

How is an intercontinental ballistic missile interpreted ecologically?

ψ

Day Seven

I WROTE QUITE A LOT YESTERDAY and, in the evening, I went for a walk down the path leading to Zen Namkhan to unwind. I was wearing a head torch, a large ring covered in cloth (folding beating sheet) I'd brought with me, and a wooden stick I'd carved with my Swiss knife. This is my equipment to search for night bugs. You put the sheet under the branches of trees and bushes and hit them with the stick so that all the bugs fall on the cloth. As I'd feared, it wasn't very interesting. I can recognise the trivial vegetation that grows in fields that have been mistreated by fire and repeated felling. The group of weevils I'm interested in is a bit fussy and demands much more natural vegetation. The odd bug did fall: a bizarre ladybird with metal colours, a pair of stick insects, longhorn beetles, grasshoppers, all sorts of bush crickets, small tree-roaches but, oddly enough, no spiders. I was a bit frustrated because although it's the ideal time, this isn't the right place for my crowd; but I was happy to have seen a few insects. We entomologists are lucky to find bugs to admire wherever we go. That's the advantage of studying the world of tiny. Max Barclay, coleopteran curator at the London National History Museum, in an admirable TED talk[44] said that we who study beetles see the world with more pixels.

The surprise came at midnight. The ear, like smell, is wise enough to distinguish what's normal, everyday and foreseeable, and demands our attention when something new occurs. This is the only possible explanation to sleeping like a log with geckos on your head that now and then go "*kah-kah-kah*", a kind of giggle that would leave a newcomer white with terror and staring at the darkness. Usual night noises are part of our memory, a peaceful balm to go to sleep with once you get used to them. In the first nights I spent here, at Zen Namkhan, I helped myself into sleep by trying to make out the beeps from the whistles, the chirping, the buzzing, gecko's giggles, frog's croaking, the beating of moths against lights or the howl of a lemur in the distance. Far more amusing than mentally counting sheep jumping over a fence.

Around four in the morning I was woken up by some uncontrolled, repeated thudding I couldn't identify. I went out onto the balcony, concerned for my belongings as I'd left them on the table. As soon as I opened the door, a huge flying thing went past me. A bat? No, too pale and noisy. An owl? Not quite as big...

Thoughts raced through my mind, a spark of hope flared joyfully. Yes, *Attacus atlas*, the biggest butterfly in the world![45] Indeed, the powerful insect came back shortly after and flew about banging against everything, including myself, deliciously slapping me on the face. It lighted on my backpack, on the floor, on my notes and eventually, on the post, allowing me to observe it at leisure. How can I express the thrill I felt for bumping into something like that! It's as if a rock-and-roll fan went out to the hotel corridor and bumped into Elvis Presley in the flesh.

A magnificent male, easily recognizable thanks to its pectinate antennae, like a double comb. It must be the size of my hand (24 cm), because the tips of the fore wings project outward and look like the head of a snake. A splendid animal!

[Human achievements]

Let's get back to the psychosphere story, the Age of Man or Anthropocene, as named by Ernst Haeckel, who also coined the term ecology. Undoubtedly, the appearance of *Homo sapiens*, with its mind and its achievements, marks a turning point. It's estimated that our species and the Neanderthal man have a common origin that dates back 600,000 years, although our oldest fossils, found in Ethiopia are not more than 200,000 years old. Civilization, as a cultural organization of society, seems to have started after the last Ice Age, 20,000 years ago, when the population of *Homo sapiens* had a few hundred thousand individuals (*Homo neanderthalensis* had extinguished slightly earlier, some 24,000 years ago). At present, we are seven and a half thousand million people in the world and counting. No other mammal, whether big or small, not even colonial animals like seals and walruses, have reached figures or biomass[46] comparable to ours.

Living organisms usually depend on their own strength to solve tricky situations, but many resort to outside energy for help. Birds hover without effort; turtles use currents to travel; plants make sap come up from the roots by controlling the evaporation in the leaves which generates suction (they don't spend their own energy in this formidable task); termites regulate the temperature in their termite mounds thanks to intelligent architecture; but nothing compares to human achievement in their quest for solutions in order to progress.

The concept of progress is very subjective—especially in its social and political connotations—, but here I use it strictly in physical terms: a system progresses when it gains greater independence from the medium in which it is and, ecologically, a species progresses if its population increases the number of individuals (biomass) and occupies more space as a result of its growing detachment from drawbacks and restrictions in its surroundings. Applying both criteria, we can conclude that our species has progressed and is successful over the other species.

The cultural evolution unleashes the complexity of ideas and the speed at which they are conveyed: from the spoken word to manuscripts, on to printing, digital format not so long ago and soon, DNA molecules to be used as hard discs, without overlooking the internet, which represents a tendency to a panmixia* of information and knowledge. Everything has accelerated in parallel ever since the so called civilization started: how we use other species (agriculture, livestock, industry); the devices we have built for transport (boats, bikes, cars, submarines, airplanes...); our means of communication (carrier pigeon, telegraph, radio, laser ray, the web...); the sophistication of artificial objects (tools, toys, prosthesis...); our constructions, of unseen proportions on the planet...The Great Chinese Wall can be seen from outer space and at night, artificial lights indicate the distribution of our species and where individuals concentrate, the hubs of civilization (the great cities). A great part of the secret of our success lies on our use of the so-called exosomatic energies (outside our body) to do the job for us: fire, wind, hydraulic energy, steam engines, hydrocarbon combustion, sunlight, atomic energy... No other species has reached our level.

We can cure illnesses and repair our bodies, sometimes introducing artificial pieces or taking organs from other individuals (transplants); we can manipulate genes now, mix them up and even create what?, new species? That's right. In the past, this was exclusive to biological evolution. But biotechnology can now introduce species on the planet which have not been tested by Nature. If this is not something more than just a biosphere, may the Martian strike me dead.

---

* Etymologically, *pan* totality and *myctós* mix in Greek; that is, all with all.

Not only have we gone beyond the boundaries of the Earth with electromagnetic waves but we have also launched complex devices, satellites and space probes to other planets, and some astronauts,—with their kit of symbiotic bacteria— have travelled to the Moon or are out there going round and round.

I remember the first day I went underwater with scuba-diving gear, Jacques-Yves Cousteau's great invention. It was in the crystal-clear waters of the island of Coiba in the Pacific, surrounded by corals, fish of astonishing beauty and small black-tip fin sharks (harmless) to liven up the show. What I found most touching, apart from silence, was becoming aware of freely accessing a medium which is not ours: the water world. I imagine what Alexei Leonov must have felt, the first astronaut to go for a spin in outer space, when he watched the blue Earth suspended in the cosmos; just like him. I wish I could…

[The meaning of the mind]

Two days ago, the fifth day in this story, I figured out the meaning of life by following a reasoning which is to some extent objective and analogous to the above. It's now time to deal with the meaning of the mind but in this case, and until we know other comparable minds, it's ours we are dealing with, human beings' (*Homo sapiens*). The apparent duality mind and life, life and mind, is not such because they can't be separated, in the same way life can't be separated from the chemistry underlying it. The mind lives on life and in order to avoid religious interpretations related to the concept of soul, I like using the term amalgam metaphorically referring to the close inseparable mix of silver, mercury and other metals which, due to its new properties, is employed or used to be employed to make teeth fillings.

Human beings are an amalgam of living matter and thinking matter. The latter coincides with the brain and except for science fiction films, this organ can't be split from our body. We have reached this situation via cultural and biological co-evolution; both influence each other in a binding dialectic as it's happened with many other achievements of our species: the capacity to talk or to walk erect. Walking on our feet, for example, is something learnt from our parents; it is conveyed via the cultural channel and it has co-evolved with the change of position of the hips, which makes it possible but does not determine it. A baby chimpanzee—our closest living relative[47]—has 65% of its adult brain. A human baby has only developed 25% and will need full protection until he is 4. The later development of the brain—including the "wiring" — depends critically on the stimuli and affection he gets.

Many years ago, in villages in Mozambique I was concerned to see women spending most of the day working in the fields with their children on their backs, stuffed into a piece of cloth which they wrapped around their bodies and carried as a backpack. You could only see their heads as their arms were stuck in their bodies and they couldn't move their legs either. I ignore what level of stimuli a baby in these conditions may get, but I guess it's very low, as are the connections that should build in their brains. That's not very good. In human individuals, the cultural channel is determining, as 75% remains to be "wired".

Given our psychobiological nature, as cultural evolution has advanced, we have left aside Darwinian biological evolution, shaped by natural selection. At present, the probabilities of a human being leaving descendants with their genes—and the possible mutations they may include—depend more on their credit card than on their ability to run away from lions in the savannah by climbing trees. Medicine has also greatly distorted natural selection. In fact, it is more appropriate now to talk of cultural selection.[48]

So, what is the meaning of the mind?

The evolution of information in the Universe has opened channels to increase its complexity, to innovate and accelerate its development which, in the case of the mind, might be not quite as blind as in previous phases. Analysing the mind from the mind itself is pretentious; such task is better approached from the supersystem it is part of, the Universe, unless we assume, like in hermetism, that it's through its elements that the Universe thinks of itself, closing a cycle.

Although science deals mainly with the whys and the hows while the reasons why are usually left to metaphysics, it is not hard to establish a simple parallelism with life and find the reason for being of the mind: if the meaning of life is to live, increase the biomass and proliferate; the meaning of the mind must be to think, innovate, increase knowledge and spread culture.

[The individual and the mind]

Having followed the first part of my story with an approach—I'd like to think objective—about the meaning of life and mind as phenomena which take place in the universe we know, we should now deal with the meaning of our individual lives; the life of each one of us, because what has been set out so far leaves room enough for every individual to give a personal and conscious meaning to it, including the nonsense of not giving a damn about it. But I'll deal with this at the end.

Before that and making the most of having one more week to write, I'd like to share my thoughts about certain issues, which might catch your interest. After all, an individual's life is a jumble of experiences, learning, emulated idols and models, books read, films seen, art that has an impact on you, all sorts of contingencies, petty experiences and secrets heard; all of it on a permanent layer of instinctive impulses which come on and off with insolent cheek. In addition, feelings keep on shaking the mixture with joys and traumas, while reason—armed with logic—watches, filters, combines, advises or decides as if it were a great referee imbued with objectivity, but working in parallel as a hidden agent of morality, that great peculiarity of our species. Now and then, things occur which violently shake the whole system and a great part of its contents are reshaped; something as simple as a serious fright or a health issue.

I also have a feeling that society often launches down cultural paths that are far from our instinctive patterns as social mammals, straining them excessively and possibly leading to traumas. I'd like to write about this too.

Our understanding of the world based on interactions between revealed, artistic and scientific knowledge is thus kept in a very special melting pot, the mind, which never stops cooking them, although it slows down when we're asleep. On average, the brain, which weighs no more than 2% of our body, takes up 20% of the energy we use. This means some 20 watts per day, a lot of energy to move nothing. Not long ago I read about the Grand Loop[49] hypothesis and, if I understood it well, it puts forward that the brain runs along all the neuronal circuits again and again in a sort of total repetitive and automate scanning. This would explain the high amount of energy spent—the so-called dark energy—although such frantic activity is rather puzzling. At the space-time scale of neurology, I guess the lapse of time between scans might be equal to several hours in our lives, time enough for many reactions to occur. This way the whole system will be kept on line, without the need to switch on and off specific circuits. Should this hypothesis be confirmed, it will also end with the myth of us using a small portion of our brain, something I've always deemed absurd. It does sound interesting.

ψ ψ ψ

Before I continue and being half way through the book, I think I must celebrate with something special. Unfortunately, they don't have Bombay Sapphire, my favourite gin. They have *lao-lao*, traditional liquor. The first *lao* is pronounced in a long downward tone and it means alcohol; the second *lao* means Laotian and it's shorter and upward. That is, Laotian alcohol; I guess if you modulate it the other way round, you'll be implying the waiter is an alcoholic Laotian. I'd better order by pointing at it with my finger.

The concoction is some sort of rice whisky and I look through the bottle neck just in case there are snakes or scorpions inside it, as I've heard they usually add them to *lao-lao* because they're thought to be healing. The coast is clear, but the odd colour of the liquid makes me doubt about its fermentation and it could have some methylated spirit, and that means a terrible migraine in my case. Prudence advises not to play the Russian roulette with something like this. I've finally ordered a very cold mango and dragon fruit shake. It's delicious.

[The heart of darkness]

I'm writing in the upper pagoda—my private shelter—sitting under the blades of a large fan which helps me cope with the heat in this Calm Air Valley, which insists on honouring the name I gave it. Sometimes the spinning of the blades falls into a harmonic phenomenon and you can hear a muffled and harsh 'flap-flap-flap' which recalls the sound of helicopters in war films. *Apocalypse Now* by Francis F. Coppola comes to mind, so I look in front of me and picture a line of sinister Iroquois UH-1 helicopters looming behind the mountain tops—in the film they come close to the water, above the sea—to start shooting missiles and fire with Wagner's Ride of the Valkyries for background music, unleashing an apotheosis of horror, like every absurd slaughter.

It's a great film; hard, very hard, with a superb Marlon Brando in the role of Colonel Kurtz pondering on horror. The script is inspired by Joseph Conrad's novel *The Heart of Darkness*, which is part of a trilogy with the first one, *Youth* and *The End of the Tether*, which closes it. In the careful Spanish Valdemar edition I have, the prologue's writer presents the trilogy by Conrad as an allegory of the three ages of man as conceived by Juan Benet[50].

In *Youth*, the young Marlow goes East, to the Sea of China, and is set to enrol in an old ship, which he manages to do after many incidents. According to Benet, this is the first age of man, the thoughtless impulse which is self-nourishing and never compares. In *The Heart of Darkness*, a middle-aged Marlow accepts piloting a river boat for a commercial company and as he goes up the great Congo River, he encounters the mess caused by colonial exploitation under the auspices of king Leopold of Belgium, until he reaches cruelty and horror incarnated in the person of Mr. Kurtz, the company's manager (colonel Kurtz in Apocalypse Now). He'll now doubt about everything, meet his own inner shadows and on returning to Belgium, he'll lie to keep the *status quo* of the socially correct truth. This is the second age, maturity, when we start to understand the reasons behind things and try to justify the impulses of the youth we've already spent; we'll try to hold on to a reason in order to keep afloat. And finally, in *The End of the Tether*, Conrad introduces an old captain—called Whalley—who sails dangerous deltas and meanders with an old assistant who makes up for his poor eyesight; disappointed about everything and everyone, his only reason for living is the chance of seeing his daughter again, to whom he sends money when he can, the only duty he's kept out of habit. This is the third age, the age of void, once the impulse of youth is exhausted and the reasoning which underpinned middle age is no longer valid. Benet speaks of old age, of ravage and rejection, concluding that to be able to live peacefully, we must refuse to enter into that third phase of alienation.

We are more at ease with beliefs than with facts. But reason sometimes acts like one of those machines employed to drill paving, and it goes on and on until it turns everything upside down. We must not go in there, according to Benet...

What the heck! What if you are already in it? I found out who this literary man from Madrid was, apart from being an engineer, and I reckoned he wrote his first novel, where he speaks of the ages of man, when he was in his late thirties or forty. Surprising and shocking. I wonder what he thought when he was in retirement age.

[The light of understanding]

After going over the issue and subjecting it to a few showers, I concluded Mr Benet must have been going through his mid-life crisis. To me, there is at least, a fourth age beyond the "Great Disappointment", and it has nothing to do with years or experiences, but with approach and making the most of our qualities as mind-bearing mammals.

The understanding of things can fulfil someone's life once their biological mission (procreation) is exhausted or discarded, even before then. As we grow, the body of mammals, including its instincts, gradually gives way to the mind as the latter gathers information and knowledge. Wisdom, understood as the full positive use of knowledge and experience, gradually gains ground in daily activities. Therefore, just as there are levels of intelligence higher than sphincter control, there are also levels of wisdom, so not everyone is bound to capsize in Benet's third age. Some, in fact, enter a phase of intellectual complacency and others even assume understanding as the only purpose in their lives, like a sort of path towards "illumination". Wherever we put it, understanding seems to be a good lifesaver.

Every person is a melting pot, and there are no two alike; just like there are no two identical trees.

During my night walk before going to bed, I noticed, on the left of the gate to Zen Namkhan, something among the trees. There's a small platform on a pedestal with a little traditional-style little house. It's all skilfully painted white with golden edges. I can make out some fruit and water. It's moving to think of someone feeding the birds and making the effort to build such a delicate bird hut. I go up to it and see that besides some food in a bowl, there are many other things: small bunches of flowers, unlit candles, a little tree made of brass sheets, plastic figures, incense sticks,... now I understand, offerings to the mountain and river spirits so that nothing bad happens to the place. I look inside the little hut and it's empty. It's either a shelter for spirits whenever they decide to turn up, or the saint is missing.

Later, in Luang Prabang, I'll see many more little shrines in gardens and outside houses. I go on with my walk thinking about what I've written today and what I've just seen. Yes, I'll have to deal with this matter too.

Religions are the pending matter of humankind.

ψ

Day Eight

AS I HAD FINISHED THE FIRST PART of my task yesterday, I decided to take the morning off today and visit one of the nearby villages and the waterfalls they talk so much about. Every day I'm asked whether I've been there yet and not going feels like some kind of sacrilege. Loui took care of everything. I put my boots on, pulled down a cloth hat, and hanged my satchel bag down my side with a small water bottle, plus my binoculars and a small camera. Only a whip was missing to round off my brave explorer outfit. I accepted instead an umbrella Loui offered although I didn't think it'd be useful in a monsoon downpour.

After breakfast I went down to the rudimentary boarding plank and a man called Wang came to pick me up with a narrow and very long canoe. His lively five or six-year-old son was with him, as well as another Laotian man who was slightly taller—an acquaintance or relative of his—who took advantage of a ride already paid for (quite common in this area). People here are short and slim, sheer fibre, possibly the result of sufficient but not optimal nutrition. Wang knows a couple of words in his English, like *Good 'molning', 'watelfall', 'elephan', we go…* and little else. But we understood each other with signs, nods and smiles.

We went up the Namkhan River and on its banks I could finally see some decent trees—*Dipterocarpus?*—which can be over 40 metres tall, have buttress roots and a trunk ten times thicker than those growing around my bungalow. The tallest—which survive in the middle of the secondary forest—stand out above the rest and spread out like giant parasols. I notice the total absence of birds or mammals on the banks and it's certainly not because they lack food, there's plenty of fruit and leaves. I guess they're the main source of protein for the rural population.

[Ban Suan Village]

Some six kilometres later we landed in Ban Suan, a village on the river bank, like many others I've seen in the tropics[51]. It stands on the raised flat bank, away from floods. There are around a hundred huts and houses, most of which are evidently humble and simple. The clay streets are dotted with badly kept vegetable gardens. There's no cleaning system, no running water and no gas, but there's electricity and it's ironic to see most houses have a small or large parabolic antenna. There are smartphones.

The residents are Laotians and people of the Hmong ethnic group, easy to distinguish because the former live in huts and the latter build brick houses. Laos is a large country, around half Spain, and has very few inhabitants (6.8 million), only half of which speak Laotian. The rest is made up of many different ethnic groups and I'm told they don't understand each other's language or dialect very well. They must use their hands, like me.

I saw my first Asian elephant, at long last. Despite its scientific name (*Elephas maxima*) it happens to be smaller than the African and has a far bigger head and small ears. The animal went across the village ridden by two kids who tried to guide it in a straight line by kicking and shouting. It looks at you with its little eyes and doesn't seem to be very intelligent;

I fancied it's more stubborn than our donkeys. It has no control of its sphincters and as it walked, a string of green dung droppings the size of a melon was left behind.

Elephants have been the brute force for these country folk, who now have some machinery so they use them less and less in agriculture and forest work. The few left there are employed to give tourists a ride. What would Sancho Panza say to his dear donkey if he saw him doing such a task? Well, elephants living on the Laos-Vietnam border have a tougher life. Their last service is to be blown to pieces when they tread on an unexploded mine[52].

It's terrible to be aware of the fact that vast extensions of this beautiful country are mine fields and no one knows exactly where you could be blown up.

Firewood is piled up outside the huts. I don't see many clothes hanging out to dry; there are a couple of mopeds and general hardship but no severe poverty. Some young girls are working on the fine embroideries they use on their clothes; they do it on a board, sitting outside their huts (most of which have no windows). They lack most infrastructures. I only saw a tin-roofed shop offering canned food, grain and pasta in plastic bags, different sweet drinks, some tools, umbrellas and little else. People go about barefoot or in flip-flops. It's a simple village that lives on what they collect and produce. But they do have contact with the outside world thanks to smartphones; that's undeniable.

[Itinerant agriculture]

The cultural and socio-economic development of society is unequal and this shows in the landscape. Even an untrained eye can notice that the forests in these valleys are not a natural homogeneous green mantle. Trees with varied canopies are on the mountain tops, while the more accessible mountain slopes are divided in some sort of a plot mosaic of roughly one hectare: some are totally empty and planted with banana trees, corn or pineapple; others have been abandoned and are covered with bushes; others are similar but are dotted with some small trees and, finally, some have recovered their tree mass although it isn't as diverse and high as the ones at the mountain tops. That's why we talk about secondary forests. This pattern occurs in all the tropics where "slash-and-burn" itinerant agriculture is practised. The soil is poor due to excessive rain. The precious minerals of the ecosystem are inside the biomass, captured and stored by the vegetation. Man has learnt to release nutrients by cutting and burning the forest so that ashes fertilize the soil and the plot can be cultivated for a couple of years with fast-growing plants. They then abandon it and go to the next plot. Only at the bottom of valleys I saw cultivation that seemed permanent, mainly rice. How different from the mechanized cultivation and fields of greenhouses we see throughout Europe!

The exploited forest plot, including the soil, will recover with time, getting close to its original structure, diversity and functionality although to reach it (primary forest) will take two hundred years or more. Hence the stages one can notice in the landscape. The process is known as ecological succession and I'll deal with this concept later.

[Tad Sae Waterfalls]

The stop at Tad Sae waterfalls was longer. It's a system of different level pools of amber light-green water, with intertwining trees which give it quite a charming jungle-like atmosphere. These trees are undoubtedly more natural and older than those at Zen Namkhan. The area is prepared with wooden bridges and paths protected by bamboo fences. There's plenty of water running at this time of the year and many people visit it. You can have a swim in any of the pools and it's a real treat if you're drenched in sweat as I was. There are some stalls to eat, souvenir stalls and unfortunately, some cages with a macaque, a small gibbon, two Guinea hens and a pair of peacocks. The magic of the air persists despite the fair they've set up in it. And sure enough, it has a small herd of elephants to give tourists a ride. From their faces I figure out they come from all over the place: Japanese, Chinese, European, Indian, Australian, etc. For the enjoyment of the boldest, they've built some platforms very high up on the trees and have connected them with steel cables. Users are provided with a harness, hung from the cable and launched to the next platform. Fortunately, the clamour of the waterfalls muffles the howls of these human parcels as they go past above your head.

I might dedicate some lines to tourism in developing countries later. Like every activity whose numbers are difficult to control, it can both be a blessing for their economy and a curse on their idiosyncrasy. I've seen too many blunders, cases of ancient cultures being trivialized and perverted, so as not to feel a twinge of sadness[53]. But people look happy and that counts.

Happy?... no doubt, I am. On a bush on the side of the road I saw a beautiful click beetle nearly four centimetres long, of an intense metallic green, and two coppery strips along its elytra; the belly was also metallic-reddish. I won't say it looks like a piece of jewellery as jewels would be happy to emulate such beauty.

There's a book by Schumacher entitled *Small is beautiful* which has turned into a commonly mentioned aphorism in events—both orthodox and alternative—related to life, nature or environmental issues. I won't deny it, least of all today. Margalef used to mention it in his talks to then add sarcastically... "but big is powerful", quoting Ken Thompson. Don Ramón knew what he was talking about. Information means power not only in espionage, war conflicts and politics; wherever it concentrates, it can influence its surrounding.

[Instincts]

Oh dear! The ants have climbed back on my table. Only two this time: one of them is tiny and the other is four time its size. I'm expectant for a while but nothing happens; the small one slips through the legs of the big one, which pays no attention. I blow them to the floor.

I've enjoyed myself on other occasions watching their brawls when two rival ant colonies meet, or different ants; although it doesn't compare to the strategies and war-like behaviour—including beheadings and slavery—Mark Moffet, the Indiana Jones of entomology, has described. One of his popular works[54] is entitled *Battles among Ants resemble Human Warfare*. When I read it, I asked myself the opposite question: how much of ants is there in humans?

Changing scales, from the individual to the group, it's easy to recognise the pulse of instincts in group behaviour. There are many social animal species which form groups or great colonies because natural selection has favoured these superstructures. It is obvious that group behaviour works towards its own acknowledgement and cohesion as well as its protection, including the individual and the genes it carries. Ultimately, the group's aim is to secure a territory for its resources, defending it if necessary. It can also attack and conquer more vital space if resources are scarce or lacking. Does it sound familiar?

Human groups or tribes are no whim of the mind. Tribal behaviour gives a feeling of belonging (reaffirmation), allows the fulfilment of the individual and provides cohesion and success to their group. It is both inherited (instinct) and learnt (culture) behaviour, and it's very powerful in a social species like ours. However, intelligence level 5 humans are supposed to be able to control more than just the sphincters. We seem to be successful with basic instincts. No gentleman goes about sniffing ladies' rears or pouncing on restaurant display cabinets. But tribal instincts (social) are harder for us; perhaps because the way they work is not as obvious.

[Tribalism and civilization]

In my opinion, civilization is the taming of instincts. That's at least what I've noticed in cultural evolution; an arrow that starting off at tribalism points to civilization, although it still has a long way to go until it can rule over the tyranny of group instincts and reach peaceful co-habitation based on reason, the sense of justice (equity) and ethics (good/bad). It might be utopian as we may be able to control our tribal instincts and avoid unpleasant consequences, but I doubt we'll manage to get rid of them altogether; I also doubt that it is actually good and desirable to our species. Human behaviour will continue to have animal behaviour for a long time, unless genetic manipulation changes things.

Instincts are there, some are evident every day and others are lurking until a factor activates them. Sex, or the need to get food are so common that we even forget their power, which is quite considerable. If you stop to think for a moment, a great deal of the community you live in is structured around these two great attractors. From the design of a tie to a car grid, to any festive music, sex is hiding behind it, a fact advertising companies know full well.

The well-known birth increase (the so-called baby boom), which took place after the death toll of the great wars, wasn't planned. It just happened. Danger is what triggers both individual and collective instincts, and sometimes, it doesn't need to be real, just perceived as such. How many people stored food after the attack on the Twin Towers in New York?

The information that comes from instincts is with us, inherited, and it's expressed both consciously and unconsciously, often in the form of powerful feelings or emotions managed by what is now called emotional intelligence, which runs parallel to rational intelligence. It's the well-known dilemma reason/passion with all its possible combinations and limits—from the Asperger syndrome to the most uncouth fanaticism—without which we wouldn't be the charming species we are at all.

The linear passage from tribalism to civilization is a tightrope where we walk backward and forward like funambulists. The close family, the extended family, the village, the professional guild, the political party or the nation; these are all levels of group association where tribalism can be patent to a greater or lesser degree, just like excluding nationalisms. We can apply reason to find equity and justice at every level, or we can just follow our group instinct. That's why I believe laws are the highest expression of civilization, understanding them as rules agreed on freely by two or more parties who then abide by them voluntarily; and so are institutions, a feature of civilization and the outcome of reason. No other animal makes laws or institutions; at best they get together to collaborate (e.g. hunt) as it also happens with humans when they gather around common interests (societies, associations, etc.).

Tribalism is here an advanced form of selfishness —me first and then you; my group first and then yours—whether we like it or not. It's our rational side that must manage the cohesion it grants while at the same time control its effects in order not to damage other "tribes". The other option, biased and bellicose, would be to displace them or get rid of them to gain space and resources.

To top it all, in our species, group or tribal behaviour is further complicated when it permeates the world of ideas and beliefs. Fundamentalisms like jihadism are, from my point of view, an extreme form of tribal behaviour and as such, hinder the progress of civilization. How many wars have been waged in the name of an idea or a single god? There are hunting territories for ideas too and resources (followers) in them. We're a territorial-tribal species and therefore aggressive, like ants. Imagine a group of ants watching *Apocalypse Now*; they'd love it and take notes down ...

Fortunately, not every religion and belief sticks to the tribe of believers. Some, like Buddhism, focus on the self and do without the ego.

[Religions and superstitions]

I have at times thought that the highest rates of civilization will only be reached when we overcome religion. I say this with utmost respect for those who believe in an existence after the present one, somewhere else or here. I suspect religious beliefs started at the earliest stages of the developing intellect and are linked to the cause-effect "wiring" of our reasoning brain that leads to animism; first a local one to then gradually move towards higher and more abstract concepts. Why does so much religion persist?

My explanation is that the mind has been affected by the pulse of life which projects into the future and, on becoming aware of itself it's permeated by it and wants to persist too. It's a rebellion against not being able to be transmitted through descendants. The idea of coming to an end when living matter dies causes an uncomfortable metaphysical anxiety. Religions calm down this anxiety, provide serenity and hope, group cohesion and consolation in difficult times. They also offer deities we can request what we lack and fuel the spirituality of people who naturally have this inclination. In short, religions are useful for people even though some are terribly manipulative and castrating, threaten with punishment and terrible sufferings, thus exploiting the instincts fear sets on. In these cases, we should ask who benefits from it. The same can be said about superstitions; they're also revealed knowledge but without metaphysical anxiety. Why does superstition persist, even among highly educated people? Why is belief more comfortable than truth? My answer is they're connected to that associative wiring that's much faster at explaining and solving potentially controversial issues and doesn't require any intellectual effort. If we adopt Richard Dawkins's theory of memes, who makes an analogy between memes and a gene (DNA), superstitions would be memes transmitted through learning. If they persist in society is because cultural selection has not eliminated them. You might have experienced how hard it is to get rid of a superstition or might have observed it in someone else.

I sometimes stop to listen to the incredible cabalism of beautiful and ugly, suspect or unlucky numbers that takes place when someone on my island goes to get their lottery ticket. It's quite astonishing to hear something like that in the 21st century.

There's some complacency and mental laziness in superstitions, just like in urban legends and social myths. They're kept even though they are not true because they're useful to people, even if it's just to calm down the anxiety of having to make a decision; they're used as a guide and that's it.

You might have been warned against drinking alcohol if you're taking antibiotics because it inhibits the effects. I've read this myth was started during the Second World War at an English hospital where a brigadier introduced penicillin for soldiers who had been injured in northern Africa. It was a very scarce antibiotic at the time and they'd recycle it by extracting it from patients' urine. But these, as they got better, would celebrate drinking beer, thus increasing the volume of urine and making it harder to get the penicillin. So the brigadier convinced the generals to forbid the troop drinking alcohol while they were on antibiotics. Let's see what you do when you're next in this situation. Will you order lemonade…, just in case?

The issue of religions and superstitions would not be much of a problem if it weren't for the many undesirable consequences they can bring. How many people have died victim of superstitious treatments? How many have been burnt at the stake or killed in the self-proclaimed holy wars? Why so much stubbornness and narrow-mindedness with science? Why that habit of being redeemers without even asking? Why insist on making the other person happy by force?

Live and let live.

ψ

Day Nine

THERE ARE MORE PEOPLE in the dining-pagoda this morning. The young couple who arrived last night disappears as soon as they have breakfast. Loui told me they're Belgian and are travelling in south-east Asia, changing cities every day. I see no charm in going around like a grasshopper with a propeller in its behind; well, it's up to them.

Eating slowly is a pleasure and I, between sips of coffee, take my time in cutting a Laotian sausage to then dip it in a fried egg with a nostalgic yellow yolk. After removing my helmet or having a shower, like now, I often pass my hand over my head to smooth my hair down. On doing this, I bumped into something clinging onto my hand and, surprise! It's a praying mantis. I guess you know them as they're all over the world. This one is half as big as Europeans and I lay it on the table where it threatens me by opening its lethal arms in an "*ora pro nobis*" attitude, hence its name "praying". We look each other in the eye, I through my 2-dioptre glasses and it through its huge compound eyes above a triangular-shaped mouth crowned by labial and maxillary palps. Are you my Martian?

In science fiction films the most horrific Martians are inspired by this insect. But they do deserve some respect, what the heck! In their hexapod world they are at the summit of the super-predators, just like tigers used to reign in the surrounding forests not many years ago. They've been decimated through superstition[55]. I carefully put it on the banister leading to the garden and encourage it to go on hunting in places other than my head.

Then the three English kids having breakfast at the table in the middle of the room see the mantis, drop their mobiles and rush to see the stylised entomological butcher. Wow! A biophilic reaction, curiosity... some hope. Their father is an English teacher in Bahrain, as he explained in German, because he loves languages and wants to learn Spanish too. I don't know what the mother studied but neither of them twitched when their three kids sprang from their chairs. They kept busily looking at their tablet and smartphone. Damn! This is getting serious. I should talk about the evils of abundance and *infoxication*, but not now. The issue of human behaviour deserves additional thought.

[Reason and barbarism]

Indeed, individual and collective actions do not only follow our free will as we might think. Many actions are driven by instincts or are subject to biological or ecological processes which go unnoticed because they operate at the species level, at larger time and space scales than every day ones.

I'd like to insist on the issue of our species' belligerence, because it seems to be increasing nowadays and, in a way, it goes against the trend to advance in tolerance, compassion and other moral values that come about with civilization. There might be circumstances that unleash it, since territoriality and aggression are achievements of Evolution which have improved the efficient use of the natural resources available by distributing them, so there could be other factors involved.

First, we must take into account the scale change that takes place when the mind appears. A furious animal may inflict harm on another one within the reach of its claws, teeth, beak or sting[56] and it'd usually do it one by one. The scale of the aggression depends on the body of the aggressor and it happens at the moment the conflict occurs. Animals seldom use unfounded violence or delayed responses (vengeance).

With the use of reason, human beings have enlarged scales and aggression can go beyond the immediate moment and place. We plan and make devices to help us in our purpose, including killing. We put poisoned baits in the kitchen to kill cockroaches; we build bows and arrows to reach distant preys or enemies; we drop bombs from the air or build gas chambers to fumigate other human beings and make soap with their fat and ashes.

Fury and passionate response go as far as our fists or the weapon within reach at the moment of confusion. But we must fear cold reason when it's serving a passionate decision as it can extend the "dark side" of humans beyond our animal scale.

I have a German friend who every time they refer to Aristotle and his definition of man as a rational animal, goes berserk and bursts into an inconsolable fit. I guess he is thinking of the "rational" final solution the Nazis found to what Hitler called the Jewish problem[57]. Aristotle might have just meant we use reason; we're reasoning animals, without further ethical connotations*.

Reason or the ability to reason employing logic is a tool of the mind which does not contribute moral values, just like Science, which is based on it. It does not consider whether something is good or bad; its concern is just whether it is true (logical) or false (illogical). Reason can be used in many

---

* The word reason translates into German as *Verstand* (understanding) and as *Vernunft* (discernment), with clearly distinct meanings.

different contexts, both good and bad according to our judgement, and even to fool ourselves (psycho-pathological thinking). Moral, ethical values, are forged through our conscience and the imprint we get from the community we live in at a specific historical time, because morality evolves parallel to the *Zeitgeist* or spirit of the time, and I'd say it is improving, although very slowly. It is most convenient for reason and moral to go hand in hand; otherwise we clear the way for barbarism.

[Geographical speciation]

I will deal now with certain aspects on the shaping of biological species, necessary to understand a hypothesis that could explain the reappearance of conflicts like xenophobia and nationalism in the early 21st century. This is a potentially thorny issue.

In the evolution of species, particularly in that of animals, progressive differentiation mechanisms take part (mutations, natural selection, etc.) which, with the passing of time, end up generating two or more sister species from an initial one; in other words, segregated populations eventually become genetically incompatible and, if by chance, they mated and had offspring (hybrids), the latter would not be fertile, like in the case of mules (a hybrid between a donkey and a mare). This process of speciation can occur in different areas or "homelands" (allopatry) or in one single area (sympatry).

The first case (allopatry) happens when a geological or climatic phenomenon builds a physical or ecological barrier that separates the original population, thus preventing or limiting individual exchanges—therefore, genetic—between split subpopulations. They will each evolve independently and face the changes in their specific environment. They adapt, slowly build up small differences which make them increasingly incompatible until they reach definite segregation as two different species which can no longer mate, even if they met by chance. This type of allopatric speciation is the most usual one, with very different time periods, from thousands to millions of years. Continents split up and move. What today is India split from Africa some 100 million years ago to end up clashing against Asia 44 million years later, leading to the Himalaya mountain range as a result of the collision. The constant climate changes throughout the history of our planet have caused species to move, often in different directions, remaining more or less isolated (this is what happened to the primitive hominids). Lions, Bengal tigers, leopards, pumas, and ocelots are felines with a common ancestor and are a good example of geographical speciation in different continents.

[Races and subspecies]

In the 19th and part of the 20th centuries, natural subpopulations of the same species were called geographical races. These would have some differentiating morphological features but are half way through becoming independent species. At present, biology employs the term subspecies to replace natural races and uses the term race for domestic ones like those we know of dogs, cats, cows, pigeon, hens, etc. But we still say human races. Why?

It is normal for many species to be in the middle of their segregation phase, not concluded as yet, and are made up of several subspecies—the Asian elephant has a subspecies in India and southeast Asia; another one in Sri Lanka, another in Borneo, in Sumatra—which, given time, will end up producing independent species of elephants. This circumstance also applies to *Homo sapiens*, but in our case, there is a taboo when dealing with the issue and you risk being labelled a Nazi, racist or any other formula of social and political exclusion, as it is an awkward issue or contrary to human dignity.

Human races—so far no one has denied there being different human groups recognizable by their physical features—are geographical races originally associated to the different territories where they developed from a common ancestry. Therefore, in biological terms, they are subspecies. All the human races are inter-fertile, which explains why one single species is recognised: *Homo sapiens*. The error and confusion come from considering one of these races as the main "species" and therefore regard the rest as "subspecies" or some sort of inferior rank. The truth is the species is made up of the whole subspecies and none is above the other. I underline none is above the other because I know the opposite idea is deeply-rooted in too many people. Human races may have morphological and physiological differences (skin colour, hair texture, eye folds, number of sweat glands, etc.) because they have evolved in different ecological surroundings and have adapted to improve their chances of survival in them, but this has nothing to do with human dignity, which belongs in another sphere: the mind. The concept of superior and inferior applied to races will always be an interested assessment alien to science.

If my Martian friend, a zoology fan, had sampled the Earth some 20,000 years ago, he could have easily described a few subspecies of *Homo sapiens*[58] to then locate their distribution in a world map. No human zoologist has dared to do it—except to distinguish fossil subspecies—probably out of ethical refraining. It is also true that if the course of our species is now down a path other than biology, it may not make sense to describe subspecies which are going to merge and disappear. Or they may not...

[Secondary sympatry]

For the time being, I'll put aside the thorny issue of the different human lineages and the way we refer to them—races, ethnic groups, subspecies, etc.—to explain the second case of speciation: sympatry. It occurs when the original population of a species is segregated in one single place due to ecological or behavioural mechanisms which encourage a gradual isolation of subpopulations. For example: a species of rodent in which some individuals get used to being active during the day, while the others are active at night; or a species of weevil with such a large population that individuals spread to different plants and slowly get used to their taste. When choosing different ecological niches[59], natural selection will favour novelties (mutations, genetic combinations) best adapted to each one, and in the long term, the small differences will establish the separation of subpopulations until their reproductive isolation is reached. This type of speciation is less frequent and there are scientists who even doubt it can actually happen without there being a physical barrier to separate them.

However, I'm not interested in you learning about the cases of primary sympatry, as described above, but about the cases of secondary sympatry, as there's more to it than meets the eye. It happens when two populations which have been separated for some time (in allopatry), but not long enough to become new species, coincide again in the same space because the barrier that kept them apart disappears, or because some vector carries groups of one population to the other's area (for example, a tornado).

After meeting up again (secondary sympatry), two things may happen: (1) if the differences forged between the two populations are minimal or insufficient to allow the free mating of individuals, everything will be mixed within a short time and they will share the same genome again; the process of speciation is therefore frustrated. Or (2), when there are physical or behavioural differences, which despite not preventing mating, make it difficult or less frequent (less attractive, aversion, or rejection due to strange smell, etc.), there will be more mating between individuals of the same subpopulation (intra-racial) than between individuals of different subpopulations (inter-racial). What is most significant is that offspring of the mixed group are set to lose out, as they receive competitive pressure from both sides, are marginalized from the territory, have little access to resources, therefore fewer possibilities of having offspring. A negative selection against them takes place; they are annulated or eliminated and as a result, both subpopulations will stress their difference generation after generation, becoming more different quicker. Secondary sympatry forces genetic isolation faster and leads species being formed (subspecies) to occupy different niches. Therefore, sister species formed in secondary sympatry are usually more different morphologically and behaviourally than those which are strictly allopatric.

These processes are known in evolutionary Biology (see diagram[60]) affecting primates and other animal species. What about humans, who are no doubt incurring into secondary sympatry in many areas of the planet? With our increased ability to travel gained by taming other animals or employing technological devices, we have broken the barriers of our original geographical isolation. Are we in the phase (1) of mixing all together and stopping the process of segregation and differentiation which started some twelve thousand years ago; or are we facing a situation (2) where mixed races are on the losing side and we will end up separating more and more with time?

I've already told you our species has strayed from the Darwinian biological evolution, which prevails in every living being. If the mind and cultural evolution had not taken place, *Homo sapiens* would have been on its way to segregating into a bunch of species adapted to their specific geographical homelands and would not have been able to mate even if they'd had the chance to do it.

The factors controlling the probability of leaving descendants in modern humans are largely cultural (likes and dislikes, taboos, religions, castes, money, willingness, etc.). But it all happens without our innate mammal behaviours being fully annulled. Instincts have not had long enough evolutionary time to change, given its genetic nature, and they are still there, exposed to circumstances.

So, what's going on with our species? Will we merge in a common equal whole or will we become increasingly different?

Given the bio-cultural amalgam we are, there's no simple answer. It is true that biological racism, ghettos, ostracism towards mix-raced or fights among ethnical groups are unequivocal proof of biological pressure towards a forced separation, expressed socially; but it is also true that the social factor complements, modules and distorts the biological factor. Ideas like equality, justice, no race-discrimination, abolition of slavery, solidarity, equal human rights, or the unmatchable idea of citizenship[61] are powerful modulators which can override the biological tendency to break up.

[Cultural racism and globalization]

Racism based on biology has been overcome in many civilizations and "subspecies" are on their way to merging despite the evident morphological, colour and even smell differences. You'd be surprised to know the mixture of alleles* from different origins stored in the DNA of each one of us. Let's say that so far, we can see a clear tendency to panmixia[62].

Yet, if one analyses the manifestations of racism, one notices that apart from biological racism, there's also cultural racism—let's call it xenophobia—and they often operate jointly and are difficult to differentiate. This rejection of other strange or foreign (= *xenos*) cultures could very well be the same mechanism operating in biological evolution but

---

* See note 25 on page 202.

119

expressed in cultural evolution. They are both complex adaptive systems, as I explained in previous days, and it wouldn't be at all surprising they worked similarly.

The hypothesis is cultures are systems which work in a way comparable to biological species, are subject to cultural evolution and take part in the same selection and segregation mechanisms as those in complex adaptive systems. Let's admit the effect of biological secondary sympatry could be restrained and overcome by cultural patterns. But, what would happen if there were a secondary cultural sympatry with the same deep systemic effects? Who would overcome it? Culture itself?

When two cultures that had been made slightly different through allopatry, meet again, they may end up in a more or less accepted tolerance with the dilution of one culture in the other or with the merger of both. But if there are major cultural differences, a new outbreak of separating mechanisms is to be expected, which could even be abrupt or violent, especially if they are forced to co-exist.

Recent phenomena like tourism and especially, globalization, are testing the levels of cultural tolerance and resistance and it seems system resilience[63] has been overridden in more than one case, giving way to self-affirmation mechanisms and all kinds of rejection; that is, a rise in xenophobia / racism and nationalisms.

It could be different but it all points to humans being far from merging into a single universal civilization in an ideological panmixia. It's the desire of many but it's not very likely to happen. And if the hypothesis of secondary cultural sympatry with a type (2) outcome turned out to be true, cultures are bound to fortify, close up, and become increasingly different from the most disparate. It's the same mechanism that forces biological species to separate; but here it'd be cultures being different within the same species: ours.

Please note that the media, films and the internet sneak specific cultures in other cultural areas without being physically present. I don't know, for example, what Muslims think of the pornography that circulates freely on the internet. All this is relatively new and its most extreme expression may be yet to be seen.

If globalization, encouraged by the market and communication networks, keeps up their invasive homogenizing pressure, it might lead us to new times of barbarism, to the return to the tribe; fateful dark times for the universal citizenship many of us dream of.

[The birds which sing best]

I go over and over these thoughts. We like comparing and comparison is certainly a source of knowledge and new ideas. But it can also be used, especially when done with similar groups, with the sole purpose of reinforcing common values (of genre, guild, class or culture). How many jokes are there about the idiosyncrasy of different peoples, which start with something like: "there was a German, a Frenchman, a Briton and a Spaniard..." I guess they have their own in other regions. It's best to keep them there, as a joke, because such display of patriotic reinforcement can actually be pathetic and even dangerous.

I think that with all these pranks with nationals or ethnic groups the situation is always one of competence between winners and losers, or someone stands out because they are hilariously clumsy when solving something (in short, mockery). The number one philosophy that thrives in the West encourages competitiveness and I really must compare it to other more advance evolutionary achievements like co-operation. Will we reach intercultural co-operation before more powerful cultures sweep the rest away?

It's late and birds have decided to bid the day farewell with their singing. They remind me of a remark by writer Henry van Dyke that is just spot on: "The forest will be very quiet if only the best singing birds sang."

ψ

Day Ten

IT IS A DULL DAY, with that crushing stillness I sometimes find a bit oppressive. Fortunately, two new things have come up and that makes me happy, as the topic of racism and xenophobia I dealt with yesterday is not very stimulating.

They've repaired the pump that fills the garden pond. Now there's a jet of water coming out of a bamboo pipe forming a wide arc that falls sonorously on the pond surface and creates many waves. It's movement, after all. I close my eyes; the water gurgling causes a "ratatouille[64]" in me, a time leap back to my childhood summers on the island of La Palma, where there was also a water tank with a running tap making the same diuretic sound.

"Oh, memories, memories!"

Hume claimed people are but a collection of images and feelings that come one after the other at incredible speed, flowing and moving constantly. I've always been fascinated by the capacity and accuracy certain sounds, smells and flavours have to evoke events or feelings from the remotest past; as if one broke open a *piñata* and memories scattered about. I'm sure the same happens to you.

The second novelty today is that I can hear the unmistakable sound of a saw cutting wood in one of the bungalows they're building nearby to expand the business. It's a counterpoint when you're surrounded by the forest, which has music of its own that celebrates the joy of living. Memories from Costa Rica come to mind; but in that country power saws fell gigantic trees in the virgin jungle, while the same government, with its other hand promoted an eco-tourism policy based on nature conservation. It seemed to me somewhat hypocritical.

[Nature conservation]

There's no doubt the repercussions thinking matter is having on the biosphere are rather worrying and need to be dealt with carefully. We could even draw a parallelism with the instinct/reason conflicts occurring within ourselves. What happens to nature in a psychosphere?

Nature conservation is an activity I embarked on shortly after finishing my degree. I've been passionate about nature since I was a child, when I used to play in the country and the sea, when there was no television or the first channels were just starting to broadcast. I studied Biological Science because of that irresistible admiration and curiosity I feel for living beings, their overwhelming diversity of shapes and habits; their despotic beauty... And because when I'm surrounded by nature I feel at peace, safe, sheltered by a great womb.

In the seventies, environmental movements were emerging in Spain and, sure enough, I was captivated by the vibrant message of the Indian chief Seattle[65], when the president of the United States wanted to buy from him what today is the state of Washington. It was impossible not to surrender to his electrifying words if you were a young Marlow. My Womb!..., my Mother!..., my Gaia[66]!

My hormones have settled now and I think differently, but still agree with Chief Seattle on what he says about the white man. We invented the toilet roll and the piano but we certainly do many botched jobs; many of them for ourselves.

I could confirm that on my islands or on the many documentaries that became fashionable (deforestation in the Amazon, seal massacre, acid rain, etc.), Rabindranath Tagore's beautiful piece of advice: "Be like the tree that perfumes the axe that cuts it down", started to seem not very practical, verging on foolishness. I enrolled in an association that defended Nature where I could speak out, vent my anger, share the fit, condemn the "System", and demand change! Indeed, I was young and *enfant terrible* but not blind. I soon realized we were not going to achieve much crying for the destruction of our loved Nature and repeating the inevitable environmental litanies. We were a bunch of people who shared the same concern and our pressure, though not exactly ridiculous, was clearly not enough. Small is beautiful, but big is powerful. I learnt it empirically.

At that time, an old banana farm labourer had given me an interesting piece of advice: "a wall won't stop water for a long time; what you must do is go up to the water and lead it down the path you want". So I decided to get into the "system" and try from within. I trained in Applied Ecology, Legislation and Management Techniques and started to work as a professional conservationist. This was happening in my country at a time when terms like environment and ecology were starting to be heard in the media, at least to advertise ecological soap or any other linguistic folly. People then—an even today—couldn't differentiate ecologists from environmentalists and I had to explain constantly that the former is a professional and the latter a militant, like a sociologist and a socialist, and that not every sociologist was necessarily a socialist. They'd laugh at the witty comparison but would puzzle over the meaning...

Remember? Science is value free. Science analyses, measures and puts objective data on the table (e.g. water at 30º C), but does not provide valuations. When scientific knowledge is applied to specific ends—food production, human health, building—we enter the field of applied sciences and technology, where experience is also an important source of knowledge[67]. The stated and pursued aims in every case provide important criteria to assess data (water at 30º C is good for a shower but not to boil an egg).

In issues regarding nature conservation, like in most activities carried out by an organized society, science, technology and politics are involved. Scientists provide the essential data and basic knowledge; the conservation experts know which method to apply or try new ones; and politicians decide which topics deserve attention, approve the budget so that scientists and conservation experts can do their job and, ultimately, make decisions. If we want results, all three legs are needed to keep the stool standing.

As it happens in many other technologies, in issues related to the environment and nature conservation, you work with the best information available and you generally have to take a gamble as the risk of not making decisions is less acceptable than that of making mistakes. You can always learn from your mistakes and you can often correct them; this doesn't mean the principle of caution[68] should not be applied where specific cases demand it.

The dilemma of deciding or giving an opinion so that politicians take a decision is usually a point of friction between scientists and experts. Let's use the following caricature when faced with the building of a bridge. Imagine a very long table. A scientist is sitting at one end, an engineer at the opposite end. The scientist asks for time and money to study the situation (rainfall, soil texture, etc.) as deeply as possible to obtain certainties or levels of trust so as not to lose sleep over it. On the other hand, the engineer, usually bound to deadlines, needs a few details—height and length— to start designing a bridge, but at the risk of it failing due to not considering certain parameters. That's why between the two extreme positions a tight rope is spread out and one must walk it to reach a compromise that would allow us to gather enough details to reduce the risk of failure in a realistic time and money framework.

I've witnessed the distress of scientist friends who have been forced to give an opinion on issues not particularly complex. The usual answers, nearly as a rule, were: "it must be protected, touch nothing[69]", or "stop, this needs further study". If someone must be rebuked, it isn't them but those who asked the wrong professional.

On the other hand, I've always been surprised at the humour or downright mockery that is common in academic forums to refer to politicians, as if they were the reincarnation of all the evils of society and those of scientists in particular. They should read Epictetus and his wise warning: "When you point at someone with your finger, three fingers point at you."

I've been tempted to hang a notice outside my former Faculty of Biology announcing that "there is life beyond *peipers*[70]"; or this one: "politicians don't read *peipers* adding "… it's arrogant to expect them to do so."

[Reason and legitimacy]

Such contempt for the "political caste" fermenting in the corridors and classrooms where science is taught—not in all of them, fortunately—, can contribute to establish the conviction that scientists' opinion is the valid one because reason supports it. Gross mistake in the aristocracy of knowledge in addition to the ensuing behaviour I've had the chance to see and suffer on many occasions in my now long career.

In a democracy, reason doesn't sanction, votes do. Or to put it briefly: Parliament sanctions unreason. Something may be a mistake but it'd be our legitimate mistake. Here lies its importance to society, rather than on failure itself. With a bit of luck, there will be time to amend it. No one will deny the scientist is right, if he is, but he must convince the others if he wants them to share his opinion.

This is the sublime difference adopted by democratic societies based on the concept of citizenship, a concept that issues directly from the mind and is only present in our species. Humans can agree on rights and duties to make living together easier. We draw up laws to set goals which reflect the moral of the time and the values that must rule public policies: what's considered much or little, good or bad. There's no need to appeal to a higher authority incarnated by some kind of deity, to its earth reincarnation or its inevitable clever representatives (pharaohs, kings, shamans, imams, dictators, etc.). When I was young I smelled a rat whenever I heard a priest spinning the yarn about the principle of authority surreptitiously by saying "My son…, we're all brothers…, Our Father in heaven…"

We're all citizens, no more, no less, and democracy is our most sensible cultural invention.

[Vocation and ethics]

I clearly recall the best lesson on conservation I've ever been given. It was in Harpers Ferry, at a training centre of the National Park Service in the United States. I was doing a course on interpretative planning[71] and one afternoon, my tutor, Jerry made me sit opposite him, with no table between us, and rather seriously, asked me:

"Are you going to be a nature conservation professional?"

I pondered on such a wide perspective for a little while. I didn't think it unpleasant at all but quite a noble cause. I liked the question.

"Yes!" I answered honestly.

"Give me your hands" —he said, while he opened his arms and drew closer to me.

I opened my arms too, almost automatically and gave them to him, being face to face with my tutor while I considered some of his features and movements and started to doubt of his intentions...Well! I've fallen for it like a baby!, I thought.

He shook my right hand violently while I looked at him with wide-open eyes and watching out for any suspicious movement.

"Whenever you work on conservation with this hand..." —he said, and took a heavy blue book that was lying by him and planted it with a thud on the palm of my left hand —"never! never ever do anything without the law on the other".

Having said this, Jerry sat back on his chair looking and smiling at me.

So there I was at first, not reacting, with my arms wide open and a heavy judicial code on my left hand trying to process what he'd just said. My image of a green crusader riding in quest of dragons to rescue Lady Chlorophyll from the Dark Tower was shattered to pieces. Jerry was indeed a contingency and I'll be grateful for ever. His teaching is now with me and I try to convey it, my own way, to all those who deal with these issues and are within range.

[Environmentalism and ecofascism]

When you work in conservation it's easy to be carried away by your passion for nature and you can end up assessing situations following your own beliefs and values. This is perfectly acceptable at a personal level or if you're an environmental militant, but it is not valid if you work in the public sector. In public issues, goals, assessment criteria, what can and cannot be done, and procedures to avoid the abuse and immunity of power, are set by laws passed by the legislative power representing citizens[72]. If the laws in force are deficient or have perverse effects, even contrary to their aims, then they must be modified or replaced but never broken.

We, apart from controlling our sphincters and other primary instincts, should also watch out for our likes and dislikes for the benefit of living together based on respect for others and above all, law. However right you are, if you are not backed by law, sneaking your own feelings through the public system—environmental administration, for instance— to impose it on other citizens, and sweeping aside their rights most of the time, is a form of corruption, to put it plainly.

Environmentalism worries me, despite sympathising with it as I share the same or similar passion for nature. It worries me because I've seen it evolve over half a century, both in and out of Spain; there are all sorts. From the most idealist, who clings to their dreamy stance and won't accept anything being touched; to the more pragmatic, who gets involved in government duties in coalition with other parties but don't lose sight of their objectives and achieve some minor improvements. There are also fundamentalist ecomaniacs who want to impose their ideas whatever it takes, lying to others or to themselves if necessary, as the end justifies the means and lies are the cheapest tool. There are those who want to save your planet at all costs, their own way, without asking if you agree. Or at worst, they'd be happy to fumigate you if they had the chance. I've read very radical ecomaniacs who think human beings are a plague on the planet; who ask for an equalitarian status for all the species and of course, in their view, humans are cheating with our technologies so the first thing to do is to eradicate medicines, etc., etc.

All fascisms are nourished by a truth carved in marble that justifies their actions. Nazism found it through considering the Arian race above the rest. Imperialism through regarding other people uneducated and unworthy to rule the Earth. And it'll be easy for ecofascism. There's an arrogant mammal that threatens life on the planet and its own life too. There's massive extinction, deforestation, the hole on the ozone layer, climate change…The planet must be saved! And at the sound of trumpets the new redeemers will come to redeem us, repeating history and without having be asked to do it.

It might seem a crude scenario but I wouldn't discard it. I detect symptoms of ecofascism everywhere. At least in my immediate surroundings: teachers, students, colleagues who work in the public administration. It isn't serious, so far, but what's worrying is that many of them fall for it inadvertently and without realizing the risks involved. One or two environmental catastrophes will be enough—fear is the best lever to move the masses—and the path will be paved for the inevitable Pied Piper of Hamelin. Then things can take an ugly turn.

[Endangered planet?]

The day in 1992, when all nations gathered to consider worldwide agreements on biodiversity, climate change and other environmental issues in the Rio Summit in Brazil—under the auspices of the United Nations—I was in Bilbao giving a lecture on land planning invited by the Aranzadi Society. I can't remember which magazine it was but I do remember its front cover all over newsstands. It showed an image of the Earth from outer space with a great headline: "The Earth in danger". This is what they believed—and many people still do—except for my dear Martian.

"Come on, you're kidding me!" and he throws the magazine back to me. "Your planet has been through far more drastic shake-ups; it's been covered in ice, over 90% of the species have been extinguished, it was hit by a gigantic meteor and has endured plenty of pandemonium. And life goes on. Look at yourself!" and he points at me with his hand.

"If dinosaurs had not disappeared, you might be a tiny four-footed animal no bigger than a shrew sniffing dead leaves to find pink worms…"

This Martian is a real pain in the neck, but he isn't short of arguments. The Earth has been exclusively bacterial for most of its existence. In fact, it's bacteria that rule the main vital systems on the planet, never mind us forgetting about them. And they're a hard nut to crack.

The biosphere was set up by bacteria before multicellular organisms appeared. In addition to photosynthesis, they have more than ten metabolic types to develop biomass from inert matter; we, animals, not even one. We depend on others' biomass. There are bacteria, especially the primitive ones—they're actually a separate realm, the Archaea—which live inside ice, in geysers' boiling water, in the sulphur of volcanoes, in underwater pits, buried 2,800 m and feeding on pyrite, in brine, in acid or alkaline mediums, etc. Science classifies them as extremophile organisms (lovers of extreme) because these environments are regarded extreme in the present time, but they actually represent what our young planet must have been like when life started to develop.

If my Martian friend was an industrial spy, and he came to have a look at Earth, I'm positive he'd choose to take with him a full bacteria and archaea kit, with their entire chemical arsenal. And he might include an individual of *Homo sapiens*, as a cosmic curiosity…

There's plenty of potential to keep living matter on our planet. Should the caloric balance on Earth change sharply towards either too cold or too hot, plants and animals might disappear but there would be many bacteria holding the torch of life.

"Ok!" says an ecologist who meddles in the conversation – the planet is not in danger but the biosphere as we know it today is! And the cause of the current great extinction[73] and climate change is our species who is on its way to self-suicide [sic]."

My turn to speak:

[Impact of the mind]

Living matter appeared on Earth, altered the composition of the atmosphere, the albedo[74] of the emerged lands, created an ozone layer which protects the surface from the ultraviolet radiation of the sun, contributed minerals, removed carbon from the atmosphere which is now sequestered in the shape of rocks that were previously non-existent, natural gas, oil or mineral carbon… Global changes which have affected the whole planet, just like its evolutionary milestones: photosynthesis, lungs, predators' teeth, fowl's feathers and many other achievements have all had deep consequences on the forms of life and the biosphere as a whole, leading to the disappearance of some species and giving others the chance to thrive.

"So how,—I wonder—can we expect a very powerful achievement of Evolution like the mind, not to have repercussions on the other organisms and the planet itself?"

What would be incredible is for it not to have considerable impact. Human beings manage external energies like no other species had done before; we manage information at unprecedented speed and distance; we use endless gadgets to increase the scope of our actions, we cultivate the land and the sea, we fly the air and go into sea beds.

"What did you expect?"—on asking this question I add that very expressive gesture of Italians: putting their fingers together and shaking them in front of their nose.

We live in a psychosphere. This is a fact we must accept and I'm afraid a great deal of the environmental movement clings on to idyllic models of nature and dreams nostalgically of returning to the "splendour" of the past. The conservation policies I know are also of a biospheric type; they take the biosphere as their model and that's looking back; it's moving backwards, playing in the losing side.

The present is a psychosphere and the near future will probably go on being a psychosphere. As it has been deduced from fossil records, the time an animal species lasts varies greatly, and the average lifespan is 4 million years. The same goes for the different extinct hominids. The average life of apes is around 500,000 years. *Homo sapiens*, who's been around for some 150,000 years (250,000 at best), could hope to live between half and a million years (as *H. erectus*), but this would happen in a biosphere scenario and following natural selection. Neither of these circumstances is applicable any more. In previous pages, I told you human evolution now depends on our own cultural successes and failures. It is so, without entirely overlooking the limiting factors of the medium, despite the gradual independence we have gained by putting on warm clothes, building houses, farming, etc.

So we should change our point of view and focus our efforts on designing a psychospheric future in the most favourable way to our species, because regarding survival we must keep on being selfish. We should need to keep many resources alive although not every and each of them, despite it being an agreed on and desirable goal[75]. This said goal seems impossible for ecological and thermodynamic reasons. I mean:

When systems containing living beings, that is, ecosystems, suffer major disturbance—sharp energy inputs, felling of vegetable cover, for instance—their structure becomes simpler, and they go back to early stages of development. Once the trauma is overcome, they follow the path towards maturity with the diversity and stability this implies. But this recovery will only happen if the disturbance stops. After my visit to the village of Ban Suan a few days ago, I mentioned the slash-and-burn itinerant agriculture as a good example of biological succession, which is a case in point.

[Energy inputs]

From the moment we started to use fire to the latest nuclear reactor, humans have never stopped extracting, releasing and putting into circulation considerable amounts of energy which involve the rejuvenation of ecosystems, including the planet as a global ecosystem. Humankind has not stopped oxidizing[76] the biosphere and devouring biodiversity; and it all points to the fact we will continue to do so unless we give up much of our well-being and comfort. The efforts of governments and technological enterprises are focused on finding new sources of energy, on finding the golden dream, that endless, cheap, and clean source which can save humankind. It may be clean but if it was not in circulation before, its large-scale incorporation is sure to have effects on the global ecosystem—more species will be extinguished—and I believe this is an aspect that has not been taken into account. In short: the success of our psychosphere entails a high rate of biodiversity loss. I don't like it but I won't fool myself.

My Martian friend, the spontaneous environmentalist and myself are by the fire ruminating uneasily and engrossed in the hot ashes, as if they concealed an answer to the future.

We're actually waiting for Godot, but he'll never come because he's an entelechy. I know that full well.

[Climate change]

Twice yesterday a breeze got up and I expected the downpour. Not a drop. During our evening chat I told Loui about it and he solved the issue with:

"Well, it's the exception that confirms the rule."

This comment might be valid for daily life but in science, one single exception shatters your hypothesis and you must look for an alternative or a wider option to encompass all the casuistry. Rules only state the frequency of events, which in itself implies this frequency is broken at some point. The weather, with all its devilish mutability, is packed with these rules and if you still doubt it, go over the Spanish *cabañuelas** and refrains.

In the recent past, humans have attributed themselves the so called "climate change", quite a funny name, because if something changes on this planet, at all time and space levels, is climate.

Climate is a highly complex chaotic system which depends on solar activity, on what happens in the depths of the earth[77], on the caloric exchange between three fluids—air, water, soil—which in turn depends on the gases they contain, the colour of the substratum, the activities of the biosphere and the psychosphere plus the heat the Earth radiates to outer space.

Do we actually have models that take into account all these factors, with their inputs and outputs, which weigh up what is the role of our species in it all?

Climate change greatly depends on the carbon cycle—carbon dioxide and methane are greenhouse gases—and, for what I've read, the big figures don't quite tally as yet. A sixth

---

* Country folk weather forecasts. [Translator's note]

part of the carbon is still to be explained and the role of the oceans is not clear either.

Nonetheless, no one should argue at this stage that climate is not changing. Evidence is stubborn. What's worrying about it is that it's happening after a very long period of relative stability during which humans became sedentary and invented property. Had we kept on being just mammals, we'd have travelled following the most suitable climate to our physiology, as all the species try to do,—there are fossils of giraffes and crocodiles in the Iberian Peninsula—including ours during the last Ice Age or, more recently, when the Sahara stopped being a savannah a few thousand years ago. But plot property is not apt for a nomad life, nor can cities be folded up and carried on a camel. There is no doubt we are facing a major problem.

Margalef, in one of his books, gives some figures on the level of energy per square kilometre our species handles compared to those handled by the climate[78]. We only reach comparable values at specific peaks, in Manhattan and some other megalopolis. The real problems will come when humans control energy levels of the order of that corresponding to the difference between the energy absorbed and released by Earth. For the time being, and on this regard, humans are not playing on the climate "UEFA League"; we must be in the second division and I'm afraid in all this climate change issue, apart from good intentions, there's a lot of anthropocentrism and unmentionable spurious interests.

Careful! I am not saying we should just sit back and watch this great theatre play where living conditions for many human beings have turned unbearable; there are millions of displacements, environmental refugees and it can actually end up in tragedy in certain areas of the planet in not a very long time. As conscious beings, we have the ability to plan our activities and control the technology we create so that they don't make things worse; but we can also reason and realize that the current climate change is unlikely to have been caused by us and we therefore can't stop it. Thinking we can could be one more mistake of our ego of conceited species.

Climate change is not going to stop; the thing is not to make it worse with our activities; we should try to avoid or mitigate its effects and adapt to it. And we shouldn't rest on our laurels. This is an urgent issue.

Day Eleven

TWO BUTTERFLIES WITH SAFFRON-TIPPED WINGS are flitting about me, absorbed in their love game and oblivious to the thoughts and hardships of a human being who, pen in hand, is trying hard to put his ideas down on paper. Introspection is no easy task.

My interest in nature is intact but something has changed in my *Umweltanschaung* (view on the world) ever since I became aware of the psychosphere and what it involves. I know now my dear nature will keep on transforming unavoidably due to the irruption of the mind on the planet, and as a human being I'm neither unrelated to it nor an innocent party; neither are you. It affects me and I'm intrigued by where the intellectual evolution of the psychosphere may lead to; it's not morbid but scientific curiosity.

From the period I've happened to live in, there are things I like and things that upset me deeply. I'm unhappy by our clumsy management of natural resources, the disappearance of species, pollution, the ugliness we have created, the unjustified cruelty we treat many animals with, not to mention wars, slave dealing or pederasty which, fortunately, I've only known through the news and in fiction, never close to me (I suspect it would alter my world view). Conversely, I feel deeply touched when I see Munch's *The Howl*, read the proposal of the *Universal Declaration on Democracy*[79] or close my eyes to listen to the cello solo by Joe Hisaishi in *Okuribito*.

[Ecological footprint]

I haven't yielded to fatalism. I know that the natural world —as opposed to the anthropic one— will be very different from the one I know, quite mistreated as it is. The whole planet will lose naturalness and gain artifice. Even if the human species were wiped out, our ecological footprint would persist. The buildings and artefacts we've built will eventually be swallowed up by vegetation, erosion and geological processes: it's a matter of time. But the species that have disappeared because of us—about one thousand confirmed—will remain extinct. And those we've moved from one place to another, willingly (e.g. cultivated plants) or unwillingly (pests, etc.) have taken root and will not return to their original homeland. Nothing goes back to the past except for our nostalgia which evokes memories seasoned with beauty and safety.

I'm not angry with my species. I say it openly because in my biophilic fits, when I was a student, I would picture the island of Tenerife freed from so many people—except myself, of course—with its nature pristine and intact. Later, my profession was to teach me that to work in conservation, one must love human beings as much as animals. I was the director of a national park, mount Teide, and we certainly protect nature from people but we do so for people. This was another great lesson I learned from Jerry.

On the other hand, I'm pleased to see how morality is widening its scope of compassion and is gaining ground regarding our attitude towards other living beings, like the rejection of animal cruelty which is taking root even in places like Spain, where we were(?) particularly cruel. In our case, the merit must go to young people; in other countries it may date back to very old traditions or stem from religious beliefs.

In a psychospheric future there's room for human determination but, to which extent will we be able to change that part of the future we imagine undesirable? We should first of all overcome tribalism and learn to agree and co-operate faster. The course is well-known but there is still a long way to sail.

The advantage humans have is that we can choose. I decided to continue working in nature conservation because it fulfils me, although I still doubt whether what I do will be useful in the long run, apart from my personal satisfaction. I enjoy the way and don't think about the finish line.

[Ugliness and harmony]

My eldest daughter has a powerful rational intelligence coupled with an equally powerful emotional intelligence which, combined with the occasional setbacks of female physiology —what Woody Allen called the pre-menstrual Chernobyl— result in an explosive combination at times, adorable at other times, in constant dialectics; it's impossible to get bored with someone like that.

I got a phone call from her one day.

"Dad! This is so, so ugly, really ugly..." she was very upset.

"Hold it!"—I cut in, understanding her problem—"where are you?"

"In Madrid, I've just come out of the tube station, terrible, and I think I'm near a football ground in La Castellana..." she'd said enough for me to locate where she was and give her some instructions.

"Go down the street towards Atocha, down the left side. You'll see a flower shop. Look for a pot with a big plant. Ask for a chair and sit down opposite it. Concentrate on every leaf, the position it takes up in space, the distance between leaves, the branching off in the stem...Just for a while, and then call me again.

She didn't call me again but ten minutes later I got a brief message: "Thanks, dad".

For a very long time, Japan has had an official sacred forest network where people traditionally go to heal. I recently read Japanese researchers had confirmed the benefits these contacts—*shinrin yoku* or forest therapy—have on people who suffer from anxiety, are stressed or have mental disorders. I smile because I have more than once suggested—half seriously, half jokingly—that the national parks in my country should be handled by the Health Ministry and be maintained by the National Health Service, Psychiatry section.

I can neither conceive nor renounce to a world without nature as a reference for harmony and an expression of beauty. I don't think I'm quite as detached from my condition of mammal so as to be able to do without it and not suffer some kind of trauma. Half of the human population is said to live in cities, in great termite mounds, and that there are children, and adults, who ignore milk comes from cows. If the future we can expect from the large number of our species is like this, I'll go with the other half, the one with less cement and plenty of fields with large busty cows full of milk running about. I do hope there will be room for macroscopic nature, which fascinates a good number of people. The idea of a world of prevailing artifice and, probably, ugliness upsets me. Luckily, the oceans occupy two thirds of the surface of the planet and at least those who believe in metempsychosis could consider the option of re-incarnating in a dolphin.

I like to think that part of our future world, with a greater or lesser capacity to sustain life, more or less beautiful, will partly depend on the will of our species, despite the fact the different cultures don't run apace regarding civilization. Globalization may be the path towards unifying approaches, or it might represent the melting pot of new emerging phenomena.

It's started to rain, with no warning from the wind...

[The evils of plenty]

Animal populations, like the striped squirrel I saw on the fifth day, critically depend on the resources available, food, dens, material to build their beds or any other basic needs. When resources are scarce, squirrels are quick and watchful to take advantage of every opportunity. Their bodies don't build fat; they're sheer fibre ready to dash off at the slightest threat. There is no misery in scarcity but optimization. Everything's kept ready and efficiency is paramount. Lacking is something else altogether, something essential for the development of the individual or the group is not there; then, "Houston, we have a problem..." The system is compromised and it may collapse.

What's not quite so obvious is that plentifulness can lead to problems. But it does. I don't mean temporary abundance—as the full bloom at springtime in our latitudes—a healthy situation which serves to readjust systems, give oneself to sex and renew hierarchies (in social animals). No, I'm talking about something else: in the Canaries, we have a squirrel which was introduced from Africa not very long ago, or rabbits, which were imported by the first European settlers in the 15th century. The thing is, if there's plenty of food all the time, the population does not only increase, but so does the weight of individuals, which build up fat; they won't make an effort to find food and will be a better shelter to parasites, which can now jump from host to host, as there are more of them and they spend more time together doing nothing. These are the typical problems caused by persistent abundance. The defence mechanism relaxes and our squirrels are not fit. If a sudden or major change occurred—like tree felling—it could take their toll on them for the benefit of eagles or other predators.

I've always thought austerity is a far better policy than abundance, although the educational and literary commonplace we've heard and read since we were children always points towards the latter: the horn of plenty, the land of milk and honey, the chocolate factory...They put many carrots in front of us to stimulate our appetite and also, avarice. A pampered life, raising a hand to pick ripe fruit...

There's no avarice in animals. When squirrels—at least the American ones I've read about—collect far more acorns than they could eat in their lifetime, they're following their instincts. They collect as many as they can and bury them in different hiding places. They don't have a very good memory and this way they increase their chance of finding one of their larders when they need to. The remaining forgotten larders encourage the dispersion and germination of the heavy oak seeds which fall flat at the foot of the tree. A beautiful example of ecological coupling.

Abundance—except of health, if possible—usually brings problems in the long run. The fall of the Roman Empire and of many others which soared before the great thump, can find an explanation in the evils of plenty; the loosening of the cohesive elements that kept it erect.

In fact, the Theory of Economics has a name for this disaster: "the Dutch disease" which comes from the mess the Netherlands fell in after finding natural gas in the North Sea. Although it might seem paradoxical, the sudden badly managed increase of natural wealth can lead to a major increase of the national currency, damaging exports. Buying goods is always cheaper than making or repairing them; other goods will become more expensive to certain social classes and, in the long term, the pernicious effects will be felt. Spain experienced the same syndrome with the treasures brought in from America after the Conquest; Australia with the finding of gold or Venezuela or Mexico when they found plenty of oil. The production in the country suffers and when lean years come, it will be defenceless, with no industry to face the challenge and its disorganized economy, victims of plenty. We don't learn much from history, or so it seems.

[Doses and resilience]

Cybernetic systems are those which have feedback mechanisms so that the output has an influence on the input. Therefore, they can self-regulate. There are plenty of them in nature and in man-made devices. The electric water heater, for instance. The thermostat on the way out measures the water temperature and if it's below the fixed value, it sets on the resistance that heats up the water coming into the tank; if the temperature that comes out is higher, the resistance is switched off not to produce more heat. This is how a flow of water at the desired temperature is dynamically kept. The foundations are quite simple.

More complex systems—like ecosystems—usually have a myriad of cybernetic subsystems. Their combined capacity is called resilience and it allows the system to manage outside impacts without changing its status. I used it when I talked about cultural tolerance and tourism. The term is a bit obscure and it's also used in psychology to refer to the capacity people have to overcome trauma, although it's a restrictive use linked to a positive interpretation. In ecology, resilience is a fundamental concept.

I can go down to the boarding plank in the Namkhan River and empty the kitchen rubbish bin. Will I pollute the river?... the impact will hardly be noticed after some time. The organic remains will be dispersed, some will dilute and living organisms will take advantage of them; they process them, so the water quality parameters will keep their usual levels. It's one more "ecological service" of nature, as we now selfishly call it. But, what would happen if I dumped at the same spot the rubbish from all the restaurants in Luang Prabang, the nearest city? The resilience limits of the river would be exceeded, the system would collapse and our nose would soon notice the mess. It's a matter of dosage, as nearly everything that concerns the physiology of living systems.

I used to explain to my Ecology students on the Tourism degree the concept of doses unerringly and attracting their full attention:

Wine contains alcohol, which is a light very mischievous molecule that goes through the stomach and intestine walls very easily, penetrating our body. It's first taken to the liver which breaks it up and uses it, because it's nourishing. Drinking a glass of wine is ok, but if we increase the dose, the flow of alcohol molecules would be higher than the capacity of the liver to process them. So many of them will run intact down our blood torrent until they reach our brain, with the effects you all know. If you want to drink with no regrettable consequences, you can first eat something full of fat or oil, like a couple of fried eggs. Alcohol blends with fats so it'll go through the stomach walls far more slowly, so the dose—quantity per unit of time—will be drastically reduced and the liver will have time to do its job. However, if the idea is to get plastered, for whatever reason, you can drink more glasses of wine, but you'll have to shell out. It's faster to make gurgles with a wine glass or drink it through a straw (it works wonderfully with beer). The released alcohol reaches the nostrils and then goes on to the blood capillaries and straight on to the brain, avoiding the liver and shooting up the dose. Remember to drink plenty of water at night and on the following day. Your brain will re-hydrate and you may avoid the hangover.

[Infoxication]

I guess systems specializing in data processing must also have their tolerance limits, self-compensating mechanism and, in short, resilience. In complex man-made machines, like computers, it's easy to recognise when it's over its capacity for processing and self-regulating (if included) because we'll get an "overflow" warning or it would just crash.

The Florence or Stendhal syndrome is defined as a psychosomatic problem in people who've been overexposed to works of art and beauty which they're unable to process without getting upset (shaking, palpitations, vertigo, hallucinations, etc.); it's all very romantic although it might just be a myth or quite simply, an illusion. A friend of mine, an irredeemable bibliophile like myself, has called the "Blackwell syndrome" to what happens when you visit the Norrington Room in Blackwell's gigantic bookshop in Oxford. There are so many books that after half an hour you just can't read one more spine. Sheer saturation!

I wonder what capacity our brain has to handle information, especially now when, by any reckoning, we receive much higher doses than those the brain got when it evolved in the body of a primate, with its limited time and space scales. What effect can so much information have on the original wiring, on our emotional components, on the mind as a whole? Are there any limits? Is it a matter of dosage? Do we have enough capacity to metabolise information or are we facing a potential infoxication?

I've read a few things about disorders due to imbalance between the kind of information we receive and the necessary emotional maturity, to the detriment of the latter. The Swedish educational system was the first one to introduce computers in basic levels and also, the first one to remove them on seeing the worrying results obtained. I'm sure neurosciences are dealing with these matters and I hope my youngest daughter, who is a psychologist, will keep me posted on future progress.

For the time being, I'll stick to my intuition; every complex system usually has its resilience and the mind can't be an exception. I take it seriously because I'm concerned about what I see around me every day.

The mind-brain system seems to call for a concerted development of all its subsystems and when one is strained, another one is neglected. Therefore, interpersonal relationships are essential to a social species like ours and they demand a time to adapt and settle; its own imprinting defined by real contingencies and experiences.

[Smartphones]

It worries me to see children and young people today engrossed in their smartphones, replacing human contact for digital contact; hiding behind anonymity and avoiding games, fights and scuffles, which are all part of the basic learning every social mammal cub must undergo.

We're replacing the game between fellow beings with virtual games and I suspect this will bring unpleasant consequences in the long term. I don't think society is aware of the risks involved in new technologies. They do have a good side, and I witnessed it in Ban Suan, but there's also a dark side to them. We can't blame the tool—despite it being called smart—but we can blame the way it's used or those who irresponsibly put it in the hands of children who are unable to control its power of seduction. Mobiles seduce and not having them is also an exclusion factor from the nearest tribe. Contacts, videogames, music, films and, incidentally, advertising. It all fits in your trousers or shirt pocket, or in your handbag and it's always at hand.

When the cigarette industry came up with the idea of putting 20 cigarettes in a packet you could put in your shirt pocket, the mere fact of having it at hand, boosted sales and tobacco consumption exponentially. The result is well-known and health authorities are now trying to restrict tobacco use due to its harmful effects on health.

Information in regular doses brings habits that can lead to addiction, especially when all the channels are used (text, image and sound). Its role as social connector is even stronger stimulated by what Wagensberg calls the 'lever joy'[80].

He means the deep mental satisfaction humans receive when with very little effort, great results are obtained. This gratification or reward must work as a powerful selection factor in cultural evolution. The Science Museum in Barcelona set up a device to show the lever effect; when a child treads on one of its ends, a hippo is lifted on the other end. All the children showed a wide smile on their faces, evidencing the lever joy. It's the same pleasure a hunter gets when by pressing a trigger he sees a powerful animal collapse to the ground at thirty-metre distance; or when you hit a tennis ball with your racket. Sheer lever joy. The same you get when just by pressing a key your friend or girlfriend are with you without needing to move. Once you have that potential in your hands, if taken away you feel as if a part of you had been removed.

Experiments have been made by taking away mobiles from young people. The psychological withdrawal symptoms can be as severe as those produced by the hardest drugs society is most concerned about (heroin, cocaine).

There is a joke, perhaps black humour, regarding these experiments. One of the boys seemed not to be suffering any type of crisis whatsoever and was doing well without his mobile; he even looked happy. When they asked him about the experience, he said calmly:

"Oh, yes, everything's fine, no problem. I've got much more time; I go about school, I'm with my friends and here and there. I've also been to my folks' place. By the way, they seem to be quite nice..."

I remember getting very angry in an electronics shop managed by Indians, on my island, where I went to get a table radio. On buying it, the shop-assistant kept insisting on giving me a mobile phone; it's free, she'd say; no, thank you, I don't want it and again she'd insist on it being free and that she had to give it to me... Things didn't get worse because I raised my pitch and a supervisor noticed and came swiftly to us, put the mobile away, gave me the paid-for radio and said goodbye with a kind smile and an oriental bow. I left the shop feeling like an untouched virgin. I thought over the issue a lot. In a capitalist society no one gives you anything for free. It's crystal-clear to me today and I've tried to make my children understand, especially my son, who when he was younger was far too gullible for our time.

[Advertising]

We live in a complex society based on consuming which compels us to gather goods, encouraging abundance. There are a lot of people out there who want to sell you something, whether you need it or not, and if you don't, they'll try to make you want it in a subliminal way, passing it under the shielding walls of reason. Advertising is a very powerful device at the service of consumption, including squandering. Every morning, some brainy people gather around a table in a company to study which cord of your basic instincts they must touch (desire, fear, security) or which hidden frustrations they've detected in their market research, to offer you their *ad hoc* product, temptingly decorated with a red ribbon. Experts call it guerrilla marketing. And you go out naked into the street and are exposed to all those predators. Well, it's your jungle and you live in it. There you have the advertising bill board, lit adverts, television, pop-ups while you are surfing the net on your personal computer. It's a brutal attack, no truce, and your only defence is to be aware of it, trying to switch off your attention or being critical and selective. Watch out, mistrust and learn how money predators operate. You're the one who buys and not they who sell to you. That would be smart and you'll gain independence. A child must be warned against these things.

How frivolously we have allowed them to put a mobile in our pockets, as if it were a Trojan horse, connected to all that advertising. It's the gift parents give their children on their First Communion or when they start to go to school: "it's so I can be in touch with them, to watch them, for safety…"

Are we really aware of the double-edge sword we're putting in their hands?

I do hope educational systems in every country which haven't done it as yet—most of them, including Spain—introduce the use of new technologies in their curricula and explain what their advantages are, the risks involved and how to mitigate them; I also hope we can move about the digital world of information safely and without infoxicating ourselves.

[Technology and its toll]

In the framework of the general acceleration that is noticeable in the evolution of the cosmos—see Day Six—the irruption of technologies based on silicon and powered by electricity may involve too fast a speed-up for our emotional stability and freedom, which are based on carbon and biochemistry. Not only science fiction, but the near future perspective outlines a cyborg-type human being or downright bionic robots. What they don't venture to mention is their psychic stability.

We must assume that the cultural evolution now working at the psychosphere has its own paths and selective mechanisms (lever joy, etc.) and there will eventually be an outcome, whatever it is. There'll be many casualties along the way; or the rise in the cultural level of enough people might cool down the birth rate and lead the global population of *Homo sapiens* to sustainable levels for the biosphere resources (there is an overload of humans on Earth at present).

Knowledge will keep on growing and I do hope I can enjoy scientific, technological and artistic progress till the end of my consciousness. In the meantime, I do my best to minimize the time wasted on discarding the avalanche of useless information that reaches me or which they attempt to manipulate me with. I try to be cautious with what I put in my pocket and, in general, I follow technological progress a few steps behind. But I'm no anti-technology snob. I have a state-of-the-art computer, I have a personal website, I use Skype when I need to and I study the seagrass prairies on my island using images obtained with a satellite placed at 617 km altitude. But I'm just in one or two professional and academic networks which are growing like runaway mycelia throughout the computerised society. And I uninstalled WhatsApp[81] shortly after starting to use it. Too much of an effort to block useless messages or to filter those I might be interested in. I'm just a cautious mature person who has become very jealous of the time I have left. And just as I cling to a world with some remaining virgin nature to watch—beauty is not spent out of being admired—I'd like to keep on living in a community in which interpersonal relationships involve looking into someone's eyes, touching, and exchanging "humanine".

I don't know if you experience the same feeling as me when you're talking to someone and all of a sudden, they just leave you to answer their mobile phone. I used to feel upset for the lack of respect it shows. I now feel sad and just wait patiently for their return, as if they've gone to the toilet to manage their sphincters. To control the mobile, we people might need to go up a few steps in the level of animal intelligence.

ψ

In the afternoon, as I was thinking about these issues while cooling down in the garden pond, the three little English children turned up. They're about four and eight years old. After their first frolics and pirouettes in the water, they noticed some tiny beings moving close to the surface. After a few tries, the eldest managed to catch one of them in the hollow of his hand and recognised a trembling tadpole, "Tadpoles, tadpoles!"—he cried excitedly.

He showed it to his sisters who, fascinated, tried to catch their own tadpole and find mother frog. On seeing their enthusiasm, I went up to them and showed them the plant where the batrachian had laid its eggs—I'd been watching it for a few days—which now hang touching the surface of the water, with the remains of mucilage that wraps and protects the eggs. I knew that later, when they went up to the dining-pagoda they'll be buried in their smartphones, taking after their parents. It's unstoppable, but I choose to keep the image of the pool, the kid's eyes shining with curiosity, their empathic interest for other living beings and their overflowing joy.

"Tadpoles, tadpoles!"

Scenes like this should never stop existing.

ψ

Day Twelve

AFTER TWELVE DAYS WRITING HERE, at Zen Namkhan, I've been through their menu of traditional dishes—excellent, by the way—and I can give an informed opinion of Laotian cuisine. Basically, there are two ways of cooking dishes. The first one consists of pouring into a bowl plenty of a liquid made with diabolically hot substances[82]; rice is then added along with all kinds of finely chopped vegetables plus noodles or chicken, or pork, combining it at will to get a variety of dishes. In the second version, they first put all the latter ingredients—fried, at times—in a bowl, and the hot mix is put aside, in a smaller bowl, so you can decide how reckless you want to be. This spice issue requires some training, just as when you go to the gym; and that's despite my having just some yogurt and fruit for lunch as some kind of truce. At first you sweat a lot and then it gradually lets up.

The curious thing in both techniques is that all the products, which look fresh and are probably from organic vegetable gardens, are separate and can be recognised as you eat them. They ignore and are missing the culinary amalgam, the new flavours that come up when you cook everything together, as done in Mediterranean cuisine. I pointed this out to Lat, one of the Laotian girls, and she explained they eat all together and out of the same plates; if a person doesn't like a certain ingredient, they can just leave it aside. They won't find its flavour in the other products either.

[Ecosystems, youth and maturity]

I think I've been talking about ecosystems throughout this text and I now realize I haven't actually explained what they are, apart from a formidable and very useful concept to understand the world around us. An ecosystem is a type of system which contains living organisms. As I explained on the first day, the limits of the system can be more or less obvious depending on where we fixed them. The Earth is an ecosystem; all the oceans are an ecosystem; my island of Tenerife is an ecosystem; the Calm Air Valley is an ecosystem; the garden pond opposite me is an ecosystem; the stomach of the termite that has just fallen from the ceiling is an ecosystem carrying a bacterial community and protozoa.

As they contain living beings, ecosystems have interesting properties, like self-organization and evolving from simple, slightly diverse, very dynamic structures to increasingly complex structures with greater diversity of species and more stable. This structural and functional evolution is called ecological succession and, on day eight, when I visited the Ban Suan village, I used the case of itinerant agriculture as an example. Remember too, my comment on day ten: if we leave nature alone, ecological succession moves towards maturity; if we disturb it, it goes back to younger, simpler phases.

It's now time to take into account the role information plays in all this issue, and I believe some ideas we may conclude from it are interesting.

A young ecosystem is very energetic and dissipates a lot of heat; the activity of its elements is very fast and in general, it contains little information as heat and information do not get on very well (see example below). These young ecosystems—like the corn field on the slope opposite—are very productive and in fact, they increase their biomass considerably. Ancient man, without knowing about Ecology, noticed it and agriculture is based on it. When a field is ploughed and sown, we take the system back to its juvenile state, whose high production rate we take advantage of in the form of harvest; then it starts all over again.

Conversely, in the mature state, after building up many species (information) and structuring the space—forest floor, understorey, canopy layer, emergent layer, etc.—the ecosystem keeps as much biomass as possible in the prevailing environmental conditions (water, light and nutrients) and processes become slower. Virtually all the biomass it produces is consumed by its elements, and minerals are recycled within them. That is, in mature ecosystems a lot of biomass per energy unit is kept—it's far more efficient—exporting far less to the exterior and becoming closer. Besides, individuals of species which settle down in the mature phase live longer (e.g. a tree) than those in the juvenile phase (e.g. herbs).

This self-organization is achieved thanks to the great amount of information involved. To start with: the genome of all the species present, each of which has its own characteristics and specific abilities. A living being is organized according to the DNA it contains and is able to reproduce. An ecosystem is self-organized through the interaction of the living beings it contains but with no preconceived plan (there is no ecosystemic DNA). To a certain extent it is "autopoietic" but it does not reproduce.

The example about heat and information that came to my mind earlier comes from our own language. When someone gets quite angry, they get really heated, use very few words, repeat them often, speak loudly or very loudly and very fast (dissipative energetic system). However, when our mind is "cool", we speak more slowly, use a far richer vocabulary and more accurately (organized system).

[Coupling and exploitation]

It is very common in nature for dissipative systems to couple with organized systems. The former contributes energy and the latter, which has more information, control. Your digestive system is very dynamic (we don't stop eating), its epithelial cells are renewed within weeks and their main function is to provide constructive elements and the energy contained in the food it digests. In contrast, our brain, whose neurons are with us throughout life, (and could live longer if allowed[83]), manages a lot of information and consumes a great deal of energy which it does not generate; that is why it is coupled, as it obtains it from the digestive system, which it exploits, in a physical, non-pejorative sense of the word. This kind of coupling is common and it is clear who controls who. In the final notes I've included a small diagram[84] to help understand this idea.

The national policies for development and economic growth should not focus so much on getting and circulating more energy—whose consequences I've already mentioned— but on gaining more knowledge to get a better performance out of the energy available.

Information can influence its surroundings but remember that according to St. Matthew's principle (day four), a well-informed system can do more with the same input of information and in the same time than a poorly informed system. This could explain a few more things.

[Pareto and St. Matthew]

Italian engineer Vilfredo Pareto (Paris, 1848) was devoted to, among other activities, political economy and sociology. On studying land property in Italy, he found that 20% of the population owned 80% of the land, whereas the remaining 20% of land belonged to 80% of the population. This rate is found in many varied phenomena, from macroeconomics to daily issues. So, for example, 20% of articles generate 80% of the turnover in a shop; or when a new computer programme is designed, 80% of the effort produces 20% of the code while the remaining 80% is achieved with 20% of the effort. The 20:80 rate has been known for a long time as Pareto Principle and it is assumed, due to its empirical base, as a simple phenomenon or rule of nature, without further explanation: things just work this way.

Now, let's combine the Pareto principle and St. Matthew principle (see page 58). Pareto's 20:80 rate could reflect the logical result of the asymmetrical balance of information when interacting in coupled systems. A hypothesis to be considered: St. Matthew explains Pareto.

Should this hypothesis be true, it could explain Pareto's 20:80 distribution of wealth within a nation, the sharp North-South differences, or colonial exploitation. It might not just be an anomalous unbalance due not only to greed and the depraved behaviour of some over others, but the normal state of "infodynamic balance" of a system with more information than the other: the "North" with the know-how (brain) and the "South" with the natural resources (stomach).

Nature has nothing to do with justice but this doesn't mean we must accept situations like colonialism or slavery, especially if at this stage of civilization, we believe them to be appalling and honestly aspire to a fair distribution of wealth on the planet. What the hypothesis put forward would unveil is that if we do want to reach a state of fair distribution, we will have to force the system to a state of infodynamic imbalance and employ resources and energy to stay there. It could be achieved with a strong will—that extraordinary feature of our species—but only after grasping and assuming the challenge of the process and the determination it entails. The atmosphere of our planet is chemically unbalanced and has an anomalous presence of oxygen within it; this is because living matter keeps on issuing oxygen and removing carbon. Thinking matter would be in charge, in this case, of making an extra effort to take the 80:20 to the 50:50. It'd be a beautiful result, no doubt about it.

The other alternative that comes to mind to undo unfair Paretian situations is to equal or reduce the difference in information between the two systems. In my opinion, programmes of international co-operation that provide education and training are more interesting than those offering economic, financial or material help. But this is about a seriously massive transfer of knowledge, although I suspect St. Matthew is not quite as tolerant and we're going into Utopian realms. Luckily, dreams are for free.

[Human feelings]

I had tempura again last night and I think I had too much of the hot concoction that came with it. My stomach woke up slightly damaged and my whole digestive system—especially the final tract—could enter into a phase of chaotic turmoil: having the runs, to put it bluntly. For the time being, I've managed to keep my composure.

Ah, health! an important issue in life which won't be questioned by the mammal writing this, or by any other living being, if asked. "Health, money and love", the three things people desire, according to the popular ditty by Rodolfo Sciammarella. I'll talk about money later. Now, about love, that manifold feeling.

I suspect instincts underlie all the feelings humans develop. Emotions are neuro-physiological reactions of the nervous system and we share them at different degrees with many animals—sadness, joy, anger, fear, disgust—something you can see every day if you have a dog at home. The essential difference is that we perceive emotions consciously as a feeling, and this determines our mood. Also, when feelings are intense we call them passions and, in certain circumstances, they can influence our whole brain and dominate our behaviour, even enslaving our rational part. I believe psychologists call it "emotional sequestration".

If my Martian friend sneaked in as a passenger in a Metaphorica Airlines flight, with the pilot (feeling) in charge of the airship (brain), and the co-pilot (reason) as advisor, he'd be on the alert as to who is the commander-in-chief of the airship that day: Commander Fury, Commander Euphoria or Commander Peace.

It's hard to detect the instincts and chemistry underlying emotions because in our species, mental expressions of emotions—therefore cultural—are usually complicated by hair-splitting that outdo flowery baroque patterns or reach megalomaniac heights. There's plenty of chemistry underneath, something fans of psychedelic products and psychiatrists know well, and use the appropriate substances to boost or control emotions, especially those that can turn against the physical integrity of the individual concerned.

The human feeling that most catches my attention, apart from compassion*, is love. One only needs to go to literature and see the wide range of definitions it contains. I believe this is a case of multiple amalgam, which includes the instinct of procreation, (libido and sex), the instinct of protection, altruism that enables child upbringing, and something else that has an expansive nature and encourages the expression of love at large or very large scales, as if it had to leave traces of the scope of its amorous space (this makes it different from attachment). Territorial behaviour, perhaps? No other male animal leaves the name of its loved one painted around, or writes poems to the brink of extenuation, or builds the Taj Mahal or conquers kingdoms for her. No other female animal leaves self-control to its partner beyond the effects of dopamine, like "madly in love" women do. Not to mention when

---

* Compassion is willingness to remove sorrow and provide well-being to those who suffer.

the feeling of love focuses on unreachable elements (platonic love) or abstract concepts like literature, music or painting.

The motivation for profit (entrepreneurs), curiosity (researchers), improvement (sportspeople), power (politicians), compassion (health professionals), protection (police) are powerful psychological drives underlain by instinct which, if dominant, can direct a person's life. However, love has a touch of squandering and creativity that recalls the dispersive strategy of nature when it comes to reproduction (sperm, pollen, seeds), and I don't rule out it may have assimilated it from them. It's quite interesting we're a species who is prone to fall in love, in full tune with procreation as the meaning of life. The atypical and extraordinary sexuality[85] of our species also points in that direction.

[Hard and soft models]

Our mind is constantly building models in order to explain reality with the aim of predicting what could happen. We aim to anticipate phenomena to avoid surprises, be more independent and make progress. Uncertainty is upsetting; normality and certainty mean peace of mind.

This is what science seeks too but in a formal way. However, models are always simplifications which are unable to include all the factors at stake, especially when approaching complex realities. That's why ecology and economy are regarded as soft sciences. They have a wealth of concepts but have few solid models, especially if compared to biology, chemistry or physics, the most solid science of all[86]. The capacity to predict of ecology or economics is rather poor and there's a point in describing an economist as someone who spends half a year explaining why what he'd predicted half year earlier didn't happen.

[Money and territory]

What is money? There's nothing like it in nature. Our species started to handle money to make exchanges easier when excess produce did not always occur at the same time. I guess it must have started in Mesopotamia, where agriculture began followed by trade, or wherever the same need arose in other continents. Money, whether seeds or coins, was a value that could be stored and used when needed to acquire goods. Its evolution (cultural) has been impressive and uneven: from silver or gold coins, pearls or other desired objects for their own intrinsic value and manageable as units, to more elevated concepts like promises of payment written on a piece of paper, bank cheques, credit cards, transfers and most recently, the Bitcoin or digital currency. Money has risen so much it's no longer necessary to have an authority to back up its value with coffee, gold, diamonds or other real assets; it's enough for people to believe their money would be accepted by others; this has given way to lies and deceit, an ability practised by humans like other animal species, but which are much improved (more later). Money still handles units and it can be accumulated, but it's now a social pact, like laws, based on mutual trust.

I use money just like everyone else, but at times it has an awkward bewildering effect on me, almost of rejection. I look at a bank note, the information it contains and the power it might exercise depending where and how you give it. Just like a missile, I don't quite know how to understand it in ecological terms, or it might be my in-born rebellion against something that can dominate me. Something similar happened with cigarettes when I was a young university student and I smoked nonstop, even in the classroom, like most of my classmates at that time. One evening, I ran out of cigarettes and I actually stopped studying, got dressed and went out to get a packet of Coronas (I remember the brand). Back home, on taking out the first cigarette to light it, I just stared at it and thought about what had just happened. Is a tiny cylinder going to rule over me? And I've never smoked a cigarette again since. I smoke a pipe and I enjoy a good cigar now and then. But it's me who smokes the cigar or the pipe and not the other way round; a subtle difference between pleasure and addiction in my case.

Fortunately, I haven't reached this extreme regarding money, which has become an essential element in anthropic ecosystems as they can't be understood without it. It wouldn't be a bad idea to try to combine ecology and economy on a base common to both disciplines: information. Here's a challenge, in case someone wants to take it up.

Territory is the closest thing to money in ecology. Or let's say the behaviour of humans regarding money is very similar to that of territorial animals regarding their territory (males' duty). There's no doubt humans are territorial mammals although males don't go around peeing on posts to mark our "property"; it's more civilized to fence plots or put up a notice that reads "Private property, no trespassing." We also know the boundaries of the village or the common forest, if there were one. I'm sure that if I asked a resident of Ban Suan, they'd tell me where the hunting areas in their village are.

The thing is money, as I see it, represents basically the same as territory: it guarantees resources (food, shelter, etc.), access to females and safety, plus the many added bonuses in our species like access to service (workforce), healthcare, education, leisure, etc. wherever these activities are organized.

Poor alpha lion that has to patrol the boundaries of its territory, forcing his bladder every now and then, go hoarse with roaring all night long, apart from checking and keeping in line the young candidates to his position. In general, human communities, money and the external signs it can provide, give us social status without so much effort, apart from the status our "lionesses" can get, who have earned, the hard way, their right to behave and be equal to males. The old mammal pact—males look for resources, female looks after the brood; male protects females, female looks after injured male—has been overcome in the most civilized societies. The access of women to education and paid work—in short, to money—has struck down the biosphere pattern. Money equalizes, like the Colt 45 did in the Far West. In full citizenship there is no gender distinction.

[The market]

If we think of money from an ecological perspective, it's right to ask not only what its flows are, its impact and its deposits (memories), but who controls it. I can imagine all sorts of answers: from the groceries' owner who sells on credit, to the local bank that grants you a loan whose actual back up would make Pareto pale (3:100); from the bank consortiums to the inevitable cabals who pull the financial strings in the by no means sinister catacombs of power. And if it weren't true?

What if we have set off a system that has become more and more complex to the point it's now out of our control? If out of so much interconnected information a complex adaptive system emerged with its own features? If the market, instead of regulating prices as put forward in Keynes's time, were something other than a communicating vessels system?

The assumption is quite simple: the market—which is now global—could be an emerging system of the psychosphere supported by information in which humans, money, and goods take part as constituent elements, but it does have its own emerging properties which are still to be revealed, despite the constant efforts to understand it (for a start, it's chaotic). And it all works on electricity. Has anyone ever thought in depth what the stock exchange is? what supersystem it is part of? is it really in our hands or does it now have its own "life"?...

It'd be interesting to explore this scenario. For a start, it's a bit unsettling to think we, like a magician's apprentice, have unleashed Yen Sid's magic broom and can't control it anymore. That might be the origin of my mistrust of money. Is it not the bait that hooks you for life (mortgages) or a chaotic system especially sensitive to basic perceptions like promises, risks, lies and deceit? The thing is, money is there and the current psychosphere wouldn't work without it. It has the same function as the ATP (adenosine-tri-phosphate), the basic energetic biomolecule in cell metabolism.

[Lies and deceit]

It's funny because this is the second time I use the terms lie and deceit. Lying is telling or expressing something that is not true—or partially hiding the truth—with the aim to deceive the receiver. Many animals do this since getting the desired result by means of a deceit that doesn't involve much effort, is a selective advantage in nature. Remember the mantis that climbed to my head? When it's watching its prey it keeps still and takes on the colour of the background (mimesis) pretending to be a harmless twig. The gigantic moth *Attacus atlas*—which came back two more evenings—is known as the "snake head", because the tips of its fore wings have that shape, and it also has very convincing spots drawing the eye and mouth of this reptile*. That's its protection against potential predators. There's a beautiful syrphid fly flapping around the garden now and then with stunning yellow and black bands on its abdomen, like that of a wasp's and getting the respect these deserve; it's actually a fly, with two little wings instead of four which move so fast you can't even see them.

However, in this game of pretending to be something else, for the lie to have the desired effect, a certain balance must be kept. There can't be more flies pretending to be wasps than real dangerous wasps because after a few tries, they'd be found out and the trick wouldn't work anymore.

Economics is not my strength but I sometimes wonder whether bankers and financial engineers are embarked in this

---

* See the image on the cover of this book

kind of game, and crises—like the one we recently suffered in the West—are perhaps a consequence of having broken the said balance.

And that's it on the issue.

ψ

Day Thirteen

I FINALLY CHOSE LUANG PRABANG to spend my last two days in Laos. It's about 12 kilometres west and I'm taking my flight back home from there.

I packed my suitcase early in the morning and my clothes were all a little damp; this place is a bit of a sauna. I bid my balcony-look-out post farewell with some nostalgia. Everything I can see from it is secondary forest, but it's forest all the same. There are no artificial elements, no posts, no aerials. On the slope opposite, two trees are in bloom with typical tropical energy and are now covered in great pinkish flowers. There must be plenty of beetles. I look through my binoculars but can only see a scarlet-back flowerpecker (*Dicaeum cruentatum*), a bird. It's my farewell tip.

If we had three dials—temperature, water and nutrients—to rule nature and turned them all to the full, a hundred per cent, we'd have a majestic equatorial jungle; the largest concentration of terrestrial life per surface unit[87]. The forest around me is nearly half in the water dial because of the dry season, but I'm grateful to it because it gave me a shelter these days without depriving me of monsoon downpours.

[Happiness]

I'm writing this at the dining-pagoda where, like every morning, Lat, Tong, Mae and Noy have just arrived with huge fan-shaped brooms. They start to sweep the tables and the floor, doing away with wings and other remains of the banquets geckos enjoy every night—there are thirteen patrons, I counted them—apart from their many droppings: their white spot makes them unmistakeable; they're soft here but are parched dry in the Canaries. The mops will come next and then, a large breakfast.

This is the calm, kind, routine I've watched every morning. Happiness, so I've read, has nothing to do with the sudden rise of adrenaline you get when scoring a goal or winning the lottery; it's rather a steady state that comes from small positive moments outweighing negative, dull ones throughout the day.

This silly, this simple…

[Luang Prabang]

Laos is an inland country with no contact with the sea. It's surrounded by Myanmar (Burma), China, Vietnam, Cambodia and Thailand. Hence its age-old isolation and the fact that Europe wasn't aware of it until the last century, when it was occupied by the French. Forty-seven ethnic groups in a population of 6.8 million people are a lot, and my guess is they've formed allopatrically in isolated basins in this hilly, badly connected, country. The capital is Vientiane, with not quite a million people. It's located on the banks of the Mekong River, in the middle of the country, and close to the border with Thailand. However, the best known and most beautiful city—a widespread opinion—is Luang Prabang, which was the capital during the reign of Lang Xang (13th century) and spiritual centre of Buddhism, even today. It has some 80,000 inhabitants and it also rises on the Mekong, just where it is joined by the Namkhan River, an old friend now.

The old city—World Heritage Site since 1995—practically centres on a long peninsula less than one kilometre long, and three hundred metres wide; the tallest buildings are the temples, it's clean and quite nice.

I look for a hotel with the help of the cheerful driver who's driven me in from Zen Namkhan. The problem is there are nearly as many as temples, or more, if we add the guest houses and other options for backpack tourism. The economic drive of the city is pretty clear. He first drove me to a newly opened luxury hotel, modern and impeccable, with large rooms which are quite as good as any in 5-star hotels back home (so is their price), but with no views, closed on itself around a very well-kept garden, like a great bubble. With so much modernity and luxury I could be anywhere in the world. It'd be a giant leap and I'd have switched off from the surroundings. Not what I was looking for.

In the end, I chose a traditional hotel, the Mekong Riverview; cosier and with a post-colonial flair to it, wooden floors, views on the river and less Asian luxury, although it may sound paradoxical. The flowers decorating the foyer and my room are real, mostly orchids, and that touches my Achilles heel. A good candidate to be listed in guides of hotels with charm.

I went for a first walk, as I always do, to find my way around. Among the many temples, I was surprised to find a spa and didn't think twice about it. I went in for a good shower and a back, head and feet massage three hours long! The one on your back is similar to the Thai's, just that the merciless massage therapist doesn't disjoint you at the end by pulling on your legs and arms. Here, they twitch your skin to activate it. While having the head massage I fell sound asleep at some point and woke up with a gentle tapping of her palms on my skull. The feet massage, after the two others, was the very best ever and I discovered the jelly condition, as if my biomass was one single, even, yellowish dough, with my mind blank. When I dazed out into the blazing sun, it took me some time to get my sight used to the light, recover muscle co-ordination and be aware of the ground I was walking on. A highly recommendable chaste experience.

The hotel has a pavement bar on the other side of the street, ideal for sitting down to write. It's located right where the gentle Namkhan meets the Mekong, twice as wide, more vigorous and with muddy waters, which flow for quite a stretch without mixing with the clearer waters of the Namkhan, thus making a bi-coloured river. Fishermen canoes and other bigger ones with a roof, which are used for tourists, move about. Also, many vegetal debris float down the river, including branches and leaves. I was deeply moved by a similar parade the first time I saw it in the Amazon, at Leticia. Biologists like watching these "little floating islands" sometimes packed with stowaways, a possible way of colonization of oceanic islands when, on being launched by the rivers into the ocean, they reach beaches and deposit their biological load on them. Exciting, isn't it? To end up in the belly of a seagull that flies by...

[Humanine]

My pen is restless and I believe this is a good time to explain a word I slipped in one day, don't know which.

In 1977, the first time I travelled to the United States, to Washington DC, I was amazed by the paper towels, disposable plastic cutlery, how clean public areas were, people not being afraid of making a fool of themselves at all—which is an advantage although it might be rather unpleasant—how practical and efficient their Administration was, their splendid healthy but insipid food and other similar things, all new to me. But what shocked me most, because I only noticed it gradually, was that the people I was introduced to welcomed me with a wide smile, would repeat my name every now and then, shook my hand to greet me; a bewildering overflow of friendliness, taking into account I was a young man in my twenties. Was it really hospitality or just canned politeness? There was something missing and I soon found out what it was. In that correct aseptic treatment there was no warmth, this honest, at times nosey interest, so common in Mediterranean countries... there was a shortage of humanine.

I gave it the name of a drug because humans are a social species and we need to have contact, we need that mortar made up of interpersonal relationships which bond people together and make groups cohesive. Humans are addicted to humans. And if we go without our dose for some time, we have withdrawal symptoms. Every culture, every idiosyncrasy, is characterized by the level of humanine that circulates freely in it. In Spain and Italy, it travels generously; probably even more so in Africa. I remember the impression I got in the city of Laayounne the first time I saw Saharan men walking and holding their little fingers together or chatting in a group, milling about on the floor behind a shop counter.

I can't remember where I read a very funny story. It was about the meeting of an Egyptian diplomat with his British counterpart in a great cocktail for ambassadors from different countries. The Egyptian man, looking for trust, came so close to the Briton that he almost touched him, while the snobbish Briton would instinctively step sideways to keep his personal space; and they spent the evening dancing a waltz across the hall while they had a friendly chat.

There are places were humanine stops circulating and the symptoms are very noticeable: they invent the most varied religions to have a pretext to touch and hold hands; or they build shopping centres to restore some of the humanine that used to flow in squares which aren't there anymore or never were. I noticed it in some parts of the United States and often in many cities, great cities I've been to. Oddly enough, the more people there are, the less humanine is shared. I've felt the bite of loneliness in the middle of a city, surrounded by people so close and yet so far. In nature, despite being the only human being for many kilometres around, I've never experienced this feeling and I can enjoy the solitude it offers. Loneliness is a human thing.

[Ikigai]

At this stage of my story, I think I have more or less explained—too much ornament, perhaps—the way I understand the meaning of life and mind as cosmic phenomena, as well as the role played by our bio-cultural species in the evolution of information; all of it within the context of my own view of the world. At the end of my website[88], I summarize it as follows:

"If the meaning of life is living, increasing the biomass and proliferating, the meaning of the mind must be thinking, increasing knowledge and spreading culture. Always with joy and, if possible, harming no one. These are my basic tasks as a human being."

However, within this general framework, there are many different ways of providing the same approach with content, so that every individual can find a reason for their own life. It's obvious we aren't going to procreate every day but we can have, in cultural terms, a purpose in life, a great vocation, a leitmotiv, as we say in Europe or an *ikigai*[89], as the Japanese say. I didn't know the concept of *ikigai* and I owe it to my second daughter, who's an artist and immediately got the spirit of the word. It's something that makes you active as soon as you get up, you try to give your time to and makes you feel fulfilled when you are at it. Those who have it can count themselves lucky, as it isn't very common for people to find their own *ikigai*. And it can always be the same or change with time, successively.

Many people are adrift, like Captain Whalley in Conrad's novel, or they just live out of habit –they vegetate, in colloquial terms, without the awareness of being a psychophore individual or wondering about metaphysical questions. A large portion of people is fooled after taking up interests filtered by third parties who pursue their own aims. Motivation imported through deceit is motivation after all, and it'd keep you going in practical terms, but it's not your own motivation and it's therefore fragile. The *ikigai*, "the reason for being" is searched for within oneself; it's the most subjective thing we can conceive.

What I've said so far about the meaning of life and mind involves our species, which is a supersystem all human beings are part of. The individual meaning of life doesn't necessarily have to be the same as that of the supersystem. There are people with no descendants or who don't contribute any idea to culture. There are all sorts. The problem would only arise in the extreme scenario of all humans deciding not to reproduce; our species would become extinct.

Some focus their life purpose on themselves while others focus it on other people: humankind, in moral terms, or society as a whole. Both positions are perfectly valid in our social context. The former doesn't do it alone and they must have an effect on the others (hopefully positive); and those who help others or deal in the public spheres fulfil their own sense of duty this way, which is a very powerful *ikigai*. The wise Japanese say that people with a strong *ikigai* live longer. I ignore it, but I'm sure they live intensely and I believe in the duty to live; it comes along with the right to live.

Egotism is an absolutely normal biological stance related to the possessive and survival instincts and, in principle, shouldn't have pejorative connotations. The person you're going to spend all your life with, the first one you must love, is the one you see in the mirror every morning: yourself. The rest, your family, your group, your tribe…go to the second, third etc. planes, except for the altruism operating in the opposite direction and, in a very particular, outstanding way, during your children's upbringing. The morally reprehensible egotism is when the whole approach boils down to me: me first, me second, and then me, and just in case, me again.

Dear reader, I don't know if you're a man or a woman, or your age, or whether you've found your *ikigai* and feel fulfilled. There are heaps of self-help books with all sorts of recipes. At a book fair in Buenos Aires I saw warehouses full of titles like "Ten questions to find the meaning of life", "Become a shaman in seven days", and other nonsense like that. I guess there are all sorts, and they may be useful; I have no experience in these readings.

The way I see it, you give your life whatever meaning you wish, regardless of it fitting more or less with what I've suggested for life and mind on a global scale.

[My case]

I'll tell you my case as an example. I've lived intensely and done such different things as agriculture, teaching, public service, international advice, legislation, research, and other not so relevant activities, in addition to the ones I may choose in the future, perhaps politics or oil painting. I was educated at a German school whose training is based on order, discipline, the sense of duty and being self-sufficient (at least at that time), which gives you a good background to face the challenges of a competitive life. On the other hand, I've always followed the proverb *quae tu faciet*, "do what you are doing", adding "conscientiously"[90] and, in general, I've done well. In any of these activities I've found peace, acknowledgment and I've felt content and fulfilled most of the time. However, on second thought, my *ikigai*, in the deep sense the Japanese give it, is perhaps my dear beetles. I've never given them up despite my professional changes. They've been a sort of umbilical cord with nature, in a 1:1 scale throughout my life and I believe they'd continue to be if I had one more life. Whenever I go to the countryside to collect them; when I study them under the magnifying glass; when I draw or write about them, time flows and everything makes sense. I don't know if it's relevant or not but I've never got angry when involved in these activities.

Besides, there's something beyond the bare fascination for these diverse ever-present elements of nature, which I regard as beautiful and interesting (each to their own). I've actually studied a specific group of weevils in depth and at this stage I must be the specialist that knows the most about them in the world. Being aware you are a leaf, certainly humble, but part of the tree of knowledge is most rewarding. On the other hand, I've discovered unknown species and have given a scientific name to about a hundred of them. These names are written in Latin, according to rather strict rules, in addition to the author's surname and the year the species was made known and published. For example, *Canarobius chusyae* Machado, 1987, dedicated to my wife, who is called Chusy.

My one hundred putative children bearing my name will remain after I die, and this is a cultural way of transcending; a perennial anxiety imposed by biology which human beings carry within us, perhaps males more so than females. I've always thought pregnancy and giving birth give women an enviable advantage over the anxiety of transcending, as there is no more obvious evidence than gestating and giving birth to a baby. Males have to be happy knowing they're our children; a mental conviction, and it's not the same. Hence the search for cultural "substitutes".

[Transcendence]

In *The Human Comedy*, William Saroyan suggests that parents live on in their children, and he doesn't mean genetic material, but manners, postures, gestures, tone of speech, all that mimic inheritance, even ideas, which are conveyed down generations via the cultural channel of families.

Sculptures, works of art, the craftsperson workshop, the cultivated land, bugs discovered, the planted tree...will persist as physical memory once you've disappeared. We humans leave trace of what we created; animals only of what they were*.

How many buildings and monuments have been built with the hidden purpose of transcending? Disciples, works of art, books... are ways of leaving intellectual descendants. This very book I am now writing, is it not a kind of intellectual legacy?

Life is reproduced and projected in time. The mind is not reproduced, but it's impregnated with the pulse of life and it adopts the will to persist; or seen under a different light: it rebels against the idea of disappearing when the body of the mammal stops working. This is existential anxiety, a fertile soil for religions which, basically, promise believers transcendence, survival and a future for their conscious being (the spirit or the soul). Is it so hard to accept that our thinking matter dies at the same time as the living matter; that we will disintegrate to become part of the cycle of inert matter, and there is nothing else?...

Forgive me if it sounds as if I'm questioning your beliefs in the case of their being the opposite of what I say. Don't think too much about it as my opinion is as much a conviction as yours, and they're both equally valid to live by. I just can't be incoherent regarding my view on the world; that's all.

ψ

---

* The original quote is by Jacob Bronowski: "Every animal leaves traces of what it was; man alone leaves traces of what he created."

Day Fourteen and Last

RATHER THAN TRYING TO IGNORE IT, I might as well accept I'm in a tourist resort. I must say I'm a bit wary of tourism as it's led to many excesses at home. The dose here seems to be low and apparently bearable, though. So, I've spent the morning doing just that, being a tourist, my own way. It's not that bad.

[Parade of monks]

I woke up early and got right in time to see the Buddhist ceremony of *bai-san*. Sitting on low stools or crouching along the streets, the locals are waiting for the line of barefoot monks in their saffron tunics and shaved heads to appear. This happens every day. As they pass, people give them rice offerings, which the monks put in a large container that hangs from their side. They accept them in complete silence. Sometimes, a young monk returns little parcels to the faithful. No one talks. The line is long and I enjoy the ritual from a distance, as I noticed I was the only foreigner present, at least in the stretch of street opposite Wat Xieng Thong temple.

Later, at the hotel I spoke to one of the bellboys, who I'd been told was a monk. The truth is they are admitted to the temple at a very young age and spend a few years of instruction there; then, they leave to carry out basic studies, go to university or, more common now, work and try to learn English while they earn and contribute money to the monastery. Some go back to monk life but others don't. At the *bai-san* ceremony they're taught to give and receive; they don't seem to be short of rice.

The ritual I saw in the morning seemed honest to me but I'm afraid it may eventually turn into a circus for tourists. Tourism has the power of setting up a culture of its own, a grotesque version of the traditional one, adapted to Mr. Money. I do hope it doesn't end up like that.

After breakfast by the Mekong, I took one of the hotel bicycles and went for a ride along the four long streets in the peninsula; I stopped here and there, took some photos and bought one or two gifts as compensation for my getaway[91], although I admit I'm rather dull regarding these things. There are many shops on the main street with all kinds of souvenirs, as expected, but I go past them. When I travel, the things I bring back home are usually beautiful seeds of a tree, an odd stone or utensils used by the locals. From Barnaul, in Russia, I brought back the abacus used by the cashier (when I tried to buy it from him, he thought it was so funny he gave it to me for free); in Hanoi, I paid an astonished rickshaw one dollar for the filthy pipe he used; from Australia, I have an old battered boomerang stained with blood... So I tried to persuade the owner of the hotel to sell me one of the long-handled shoehorns that there are in the rooms. It'd have been a practical, great souvenir but it wasn't possible because the owner was away, in Vientiane.

[Morning Market]

The "morning market" starts at the Great National Museum along a lengthy narrow street including some side streets. Like every local market, it's a good portrayal of the country. Luckily, they don't import much and local products can be easily seen, as the invasion of plastic, vacuum packaging or labelling are not here yet—nor is hygiene, I'm afraid—although everything looks fresh. You can buy live eels, crabs clawing at the bucket, dancing butterfly chrysalides just out of their cocoon, a bundle of hens with its cock...

There are fascinating fish, some with no rear fin, others with whiskers, which they cut open right there. Tropical fruits unknown in Europe, like pitaya or dragon fruit, which reminds me of prickly pears; sweetest longam; mangosteen, even sweeter; rambutan, whose long hair reminds me of the fruit of our chestnuts; or the stinky durian, hated by some, passionately loved by others. Obviously, there's an amazing array of all kinds of pepper and such a huge and varied amount of vegetables to fulfil the fantasies of the most demanding vegans. All this despite the monsoon season and the rivers being high. During the dry season —and I've seen it in Google Earth—the banks of the Mekong are exposed as the water levels go down. Their fertility is renewed with the deposited mud, turning them into ideal farming land; the same goes for the great bars or islands that form in the middle of the river. That's what Laotians do; the first to arrive plants their vegetables and others follow suit. For half a year, the river turns into an admirable vegetable garden.

The diversity offered by a local market is like a siren's alluring call. In the past, in analogical times, you'd use all 36 stills in your roll. Now, with digital cameras, the walk in the market can take for ever, so you must put a stop at some point.

I also visited some of the great Buddhist temples still in use (over 50), just to comply. Sacred art is not my thing but I was surprised at the beauty of the carved panels and the details of the little figures in them, many of which tell the life of Buddha. At least, Buddhism is a religion with interesting ideas, especially when compared to Catholicism, which is centred on guilt and, to start off with, they burden you with an original sin.

I spent the rest of the morning in the hotel porch, reading a book I'd brought with me. It's written by Lorenzo Luengo, and we'll hear a lot of him. Its title is *El dios de nuestro siglo* [*The god of our century*], which turns out to be lies (there you are). It's written in an elaborate style that's as accurate as a scalpel, perhaps a bit obscure when explaining the psychological turmoil of the characters. His base is morbid curiosity, like every thriller, edging brutality, but you just can't put it down, even if it makes your stomach turn. The world it portrays, sordid to the extreme, is the antithesis of everything I've experienced these days and, in a way, it's offered a counterpoint.

[Physical recapitulation]

Goodbye, Laos! I've enjoyed this interval devoted to writing, reading and resting. It was a good choice and the locals haven't disappointed me at all. The Danish use the word *hygge* to describe this kind of feeling: feeling welcomed, comfortable, protected in a place where you settle and that reaffirms you, without much fuss.

Beer in hand, watching the waters of the Mekong running down, I owe my last thought to you, who have been with me without being aware of it. I'm grateful because of it, even though I don't know you, and I also feel I owe you an additional explanation if you happen to be a scientist. I only realized this a few minutes ago.

As I was walking down the promenade that runs along the river, I was startled by a thud I heard right behind me. I was motionless with fear. Ideas flowed fast but I discarded a security issue. In Laos I've only seen two very polite policemen and you don't need to cross the street unless you want to risk it, as it happens in countries of the real Third World. I turned around and saw a beautiful coconut on the pavement. It fell from very high up and I was lucky by just a few steps. I now smile and wonder what would have happened to physics had Newton been resting under a coconut tree, instead of an apple falling on his head.

I've spoken of reality and I'm aware it's an assumption questioned by physics. But my tale is Newtonian just like the jar of beer in front of me. My size and time are Newtonian scales and so is the world I inhabit and perceive with my senses. I don't have to force space-time to the limits of the speed of light to be able to go on writing; nor do I have to distinguish between gravitational and inertial mass. None of the issues I've put forward deny the quantum and relativist view of the Universe but I rely on Newton to communicate with other mortals like myself[92].

Pedestrian intuition and logic suggest there must be one physical reality. Not being able to know the quantum state of an element because we alter it when we look at it, does not necessarily imply it doesn't have a specific one, its own physical reality, even though it's unknown and not knowable. This uncertainty forces scientists to work with probabilities, to admit many parallel universes and get into all kinds of mess — as it must be, if science aims to explain reality — which we don't need for the time being in our daily life or to find a meaning to life.

Having come this far, I believe the honest thing to do — to wind up — is to briefly summarize my physical view on the nature of things, thoughts and beliefs, because the meaning of life and mind I've tried to explain here is based on them.

I've chosen to resort to a text I wrote to my eldest daughter — who studied physics — whose internet discussions with an Indian colleague surprised me with their pseudo-scientific or spiritual content when approaching issues science doesn't deal with, or only with bland hypothesis. I notice in young people — the ones I know, logically — an urge for the "alternative", be it mystical or not, perhaps out of weariness with the materialism of the West, their repulse to so much public corruption or sheer bellyful of non-processed information. I wish one could poke a finger in their mouth and vomit all that indigestible information we get landed with. It's obvious there's something that doesn't fulfil them, doesn't satisfy them and makes them feel sick. The truth is I can't think of a better way to spur them into exploring other avenues of knowledge.

So, in no time I started drawing up a list of sentences written from the most furious rationalism and sent it to her to give her food for thought. I believe they've widened their debate circle and she'll let me know if it's useful. Now I'd like to do the same with you. After all, in a way, this list of sentences sums up my physical view on the world and it serves as recapitulation and closing of this book. However, as I haven't written it during my stay at Zen Namkhan, I enclose it as an appendix after the acknowledgements.

[Bangkok Airport]

I add these lines I wrote at Bangkok International Airport on my way back to Tenerife. The stopover lasted for a few hours and I spent them resting and reading in one of the several Lounge Clubs. Some fifty million passengers use this airport every year. The main terminal is H-shaped; the parallel sections are one kilometre long each and the transversal one is one and a half kilometres long. Another world, no doubt.

On my right, surrounded by the futuristic structure made of aluminium and glass, there's a courtyard carpeted with green artificial lawn, bordered by pine trees and palm trees (real ones) with a black monticule –which I don't know what it is or what it is meant to be–, and a flat surface painted blue like a lake. Over this setting, there are about fifty carved birds painted white, (seagulls?) hovering in different flying positions, or standing on the grass or the water. It looks like frozen action, like a still photography. It might be a coarse flower arrangement emulating what there was here before the airport was built, but it's disturbing, and I'd rather think it's a sculpture through which the artist wants to convey something…, probably nature we are leaving behind? Because his work is placed in a very specific artificial context and it's somewhat grotesque.

Art is immense.

ψ

Epilogue

ON RETURNING TO TENERIFE, where I'm writing these lines, I typed a fair copy of the manuscript I brought from Laos. I've tried to keep to the original text and have only polished certain forms and phrases; I've also rearranged the odd paragraphs, giving a better explanation when necessary and suppressed redundancies. Obviously, I added the quotes and numbers I didn't know by heart and, above all, I've included most of the notes at the end of the book to give complementary information without damaging the assumed freshness of a hand-written text. Finally, I introduced labels in square brackets (not really section headings) to facilitate locating the contents.

I've worked cautiously, because it often happens that on editing a manuscript you end up betraying the original writing, but I must also take into account the book is not at my service but the other way around. And that's a responsibility. Every book is an achievement and has an entity of its own. From the author, the friends who help to outline it, the typesetter, the publisher, and the translator, we're all at the service of that human production which is both so humble and so huge: a book, launched into the psychosphere to have a life of its own.

Coincidences of life: I was celebrating my return with two good friends and one of them told me that Edward O. Wilson—I've enjoyed and learnt much from many of his books—had published the previous year a book entitled *The Meaning of Human Existence*. It's evident ideas hang in the air. I've already bought it but I've refrained from reading it until I finish my own book. Being the author of *Biophilia* (1984) I think I can guess what he'd be getting at. But I won't expose myself to an outside influence of his calibre so I can be loyal to the fourteen days I'd agreed with myself[93].

Within me, there are bits of Margalef, Wagensberg, Gell-Man, Darwin, Margulis, Marvin Harris, Epictetus, Confucius, Savater, Joyce, Woody Allen, Borges, Papini, Saint-Exupéry, Groucho Marx and many others who make up my personal set of ideas. If anything put forward in this book, whether my own contribution or borrowed from them, passes on to you or makes you bake your own ideas, I believe it will have fulfilled its goal.

Nevertheless, this book has been written with optimistic vitality and I'm aware, especially after reading Luengo's novel, that in this world there's another one that's really ugly and unfair and many people are trapped in it or have not had the chance to know a better one. If by any chance you're one of these people, I feel that, in a way, I must apologize because much of what is said here might sound to you as absolutely frivolous and tactless. I hope we have the chance to have a drink together.

And now, I say goodbye, like in a letter, with a hug and wishing you best luck with your *ikigai*.

Always yours,
Antonio Machado Carrillo

Acknowledgements

This book is a mixture of travel book, philosophical essay, intellectual legacy and popular science. Several people have contributed to it, in addition to the kind Martian who was happy to be metaphorically employed to explain scale issues.

I will start by thanking those who agreed to be part of the "testing ground" and read the first draft, sharing with me their opinion, doubts and encouragement. They include my children, Laura, Elena, Guillermo, and Eva, plus a most persistent critic, their mother, Chusy Hernández Hernández, I expected no less of her. And I'm also grateful to Agustín Aguiar Clavijo, Ana Mª Lezcano Fuente, Antonio Hernández Cabrera, Concepción de León García, my sister Cristina Machado Carrillo, Gema de la Rosa Medina, Imeldo Bello García, Isidoro Sánchez García, José Andrés Sevilla Hernández, José Luis Martín Esquivel, Juan Bosco González Delgado, Juan Luis Rodríguez Luengo, Julia Gómez Dasilva, Marta González Carballo, Ninoska Adern Febles, Pilar Romero Mur, Rafael García Becerra, and Soledad Sopranis de Olano.

The English version of this book also received the valuable scrutiny of wonderful people like Ted Trzyna, Manuel M. Mota, and Jonathan Broadbridge.

I also feel in debt with Hamza Hambali, Loui and the staff at Zen Namkhan Boutique Resort, who made my stay in Laos so pleasant.

Lastly, my special thanks to Ana Lima, who carefully translated this book into English.

ψ

Appendix

I

- Energy, matter, time and information started with the so-called Big Bang.
- If there is existence, there is time; and vice versa.
- What existed before the Big Bang (energy? information?), we don't know and can't know.
- There may be other universes, but we don't know.
- The (our) universe started with a first asymmetry that introduced a variational principle.
- The variational principle may be related to temperature or information; we don't know.
- It seems that the universe evolves towards complexity, gaining information and getting colder.
- In any exchange of energy and matter, there is a change of information.
- Information (just like entropy) is probably a property or state descriptor of matter.
- Information, related to the form of matter, can be structural or functional.
- Functional information has an effect on other informed systems that are capable to process it.

- More informed systems can do more with small input of information, than less informed ones.
- The increase in the size of things could be related to a better performance in informational processes.
- Complexity relates to the level and kind of interactions between components of a given system.
- The evolution of information towards complexity (up to communication) is a universal trend.
- Accumulated (historical) information favours exchange of energy and matter towards progress.
- A system progresses when it gains more independence from the external environment.
- Persistence is a trend in complex adaptive systems (the more informed the system is, the stronger it is).
- Matter (informed) evolves more rapidly when complex adaptive systems build up.

II

- Life (living matter) is an emergent property of inert matter and can be distinguished by its behaviour.
- Living systems are dissipative, self-maintaining, and mnemonic. They are fostered by autocatalytic chemistry.
- Life evolves openly in a contingent universe and tends to expand as much as it can.
- Biodiversity is a consequence of the process of life and contingencies (no determinism at all).
- Genetic mutations are not directed. Hence, there is no goal or purpose in evolution (no teleology).

III

- The mind (thinking matter) is an emergent property of life and can be distinguished by its behaviour.
- The emergence of mind may be a consequence of chaos or of the progressive evolution of matter.
- The mind manages highly structured information and permits communication and its propagation.

- The mind is a complex adaptive system that fostered cultural evolution.
- Cultural evolution is much faster than ordinary biological evolution which follows the genetic way.
- With the mind, determinism (willpower) started in our known Universe.
- With the mind, consciousness started in our known Universe.
- Let us use the term natural for things that exist prior to mankind or without its intervention
- Let us term the effects of the mind as anthropic, and their products as artificial.
- Our species evolves (in the natural and artificial environment) but not in the Darwinian way anymore.
- In planet Earth we have at present inert matter, living matter and thinking matter.
- In planet Earth we have a hydrosphere, a lithosphere, an atmosphere a biosphere, and a psychosphere.
- The psychosphere has expanded outside the physical limits of the other spheres of planet Earth.

IV

- As observer, we are bearers of knowledge.
- The observer influences the observation. Hence, observations are subjective.
- Observations can be real or illusory.
- Real is what has a physical entity (measurable). Hence, reality is supported by matter.
- Illusory is what is seen as real but is not.
- Our observations and deductions tend to be anthropocentric or anthropomorphic.
- Knowledge is an ensemble of mental representations (ideas) of things and events, present or past.
- An idea is real; what it represents may or may not. Ideas have a material (neurochemical) basis.

- The senses and accumulated experience (historical knowledge) play a role in observations.
- Deduced and inferred 'reality' may be real or illusory.
- Artistic (passion) knowledge is subjective, not dialectic and linked to instinct, intuition or creativity.
- Revealed knowledge is external, objective and also not dialectic (never tested, only interpreted).
- Scientific knowledge seeks objectivity, is intelligible and highly dialectic (submitted to proof).
- We use these three basic kinds of knowledge to move around in our lives.

V

- The basics of rational capacity are 'wired' on causal-effect procedures of the supporting brain.
- We build abstract ideas with our rational capacity to form coherent interpretations of gained knowledge
- We firstly seek and gain knowledge from ourselves and the environment, for moving safely around.
- We also use our rational capacity for seeking complacency and for avoiding pain, fear, and anxiety.
- With the emergence of the mind, our species is the only one conscious of its timed persistence.
- The idea of gods or an elevated unique deity is related to the original cause-effect wiring of our brain and with the protective imprinting of our dependant childhood.
- The use of religious beliefs is strongly rewarded by the complacency and calmness obtained.
- Love, universal consciousness, fraternity, and feelings alike are conceptions for the complacency.
- If there are other forms of life after the present form of life, we don't know (scientifically).
- The mind is free to think without limits (as far as we know).
- The mind is free to believe anything as long as it is maintained in the realm of revealed knowledge.

Final Notes

1. The perfidious Albion is none other than Great Britain and thus, in a hostile, almost war-like tone I have seen writer and academician Arturo Pérez Reverte use it. The occasional arrogance of the British and the unquestioned pre-eminence their language has gained make him a bit angry too. The phrase was introduced by a French diplomat and poet, Agustin L. M. de Ximénès, when he suggested invading the United Kingdom. Albion is a reference to the white cliffs (*albus* in Latin) in Dover, which is the first thing one sees as the island is approached from France.
2. Wagensberg, J. (1985). *Ideas sobre la complejidad del mundo* [*Ideas on the complexity of the world*]. Barcelona: Tusquets Editores (Metatemas 9), 154 pp.
3. In 19th century Spain and other Catholic countries, making a children's festival on 5 January was established. That night, the three Magi—Melchior, Caspar and Balthazar—bring gifts to the children in memory of those they brought to Jesus Christ shortly after he was born, according to Christian tradition. It is actually an imitation of what is done in other countries at Christmas to honour Santa Claus (Father Christmas).
4. Hawking, S.W. A Brief History of Time. From the Big Bang to Black Holes. 1988 London: Bantham Books, 256 pp.
5. *Perenquén* is the common name given in Tenerife to reptiles of the Gekkonidae family, called *salamanquesa* in mainland Spain. The English word 'gecko' apparently comes from Indonesia.

6. Ramón Margalef López (Barcelona, 1919-2004) was the first Professor in Ecology in Spain. Even though I never attended his classes, I consider myself an ingrained "Margalefian" because of the influence some of his books had on my training, including *Ecología* [*Ecology*] (1974), *Planeta azul y planeta verde* [*Blue Planet and Green Planet*] (1980), *La biosfera, entre la termodinámica y el juego* [*The Biosphere between Thermodynamics and Game*] (1982), *Teoría de los sistemas ecológicos* [*Theory of Ecological Systems*] (1991) or his excellent last work *Our biosphere* (1997).
7. In physics the adjective Newtonian refers to the classical mechanics developed by Isaac Newton. I use it here in contrast with the relativist mechanics developed by Einstein, which is more universal.
8. I have taken these figures from Sender, R. et al. (2016). Revised estimates for the number of human and bacteria cells in the body. *PLoS Biol.* 14 (8): e1002533.
9. We owe the concept of symbiogenesis to Lynn Margulis. It refers to the result of the permanent symbiosis of a cell within another cell where, in the long term, part or all of the genetic material of the tenant is transferred to the host, resulting in a new organism which blends both symbionts. The postulate is that the process could have been started by an absorbed bacterium which was not fully digested, but which was useful in some way to the cell that gobbled it and vice versa. Plant cells have at least five genomes from different origins, while animal cells have four.
10. In medieval Spain there were blood purity laws ruling military, ecclesiastic, and social hierarchy based on ancestry in order to avoid the presence of people of Muslim or Jewish descent. The Inquisition supervised the system checking who were Old Christians of pure blood.
11. Some 1,300-1,600 litres per square metre and year is the rainfall during the monsoon (May-October); the rest of the year is dry, which explains why most trees are deciduous (80%); that is, their leaves fall and so they are bare, like the beeches and oaks in Central Europe. Here due to lack of water; there, because of the cold. In Canaries laurel forests or *laurisilva*, there is low rainfall

(600–1,000 litres/m²/day) but during the dry season (May–October), which is opposite to here, the summer, trade winds blow regularly and the trees, which grow at the altitude of the clouds, capture the little water drops with their frond and let them roll down to the ground, or just drop them so that wetness is kept. That's why *laurisilva* is evergreen, as it happens in other cloudy or foggy forests.

12. In contrast with the alphabet (characters) humans use to represent the thirty basic spoken sounds, the genetic codes employ just four nitrogenous bases: adenine (A), cytosine (C), guanine (G) and thymine (T) in the DNA, plus uracil (U) which replaces thymine in the RNA. Genetic phrases are written with five letters, grouped in three-letter words. The binary code used in computer programmes (*software*) is even simpler and there are only ones and zeros.

13. Biodiversity is a contraction for "biological diversity" and it was first used in English in 1986 at the Biodiversity National Forum held in Washington DC under the auspices of the National Academy of Sciences and the Smithsonian Institution. It was Professor Edward O. Wilson from Harvard University who edited the results of the forum in a book entitled *BioDiversity* (1988). That is why he is usually believed to have coined the term, even though he recognises it was Walter G. Rosen, the official of the National Science Academy in charge of organizing the event, who came up with the successful word.

14. The formula introduced by R. Margalef (1989) in a footnote in his book *La Biosfera entre la termodinámica y el juego*, is only apt for scientists but I include it here because I think its importance has not been acknowledged: $\Delta G = \Delta H - (T/I) \Delta S$; where ΔG = free or usable energy, ΔH = enthalpy or a priori available energy, T = absolute temperature, I = Information and ΔS = entropy. He also puts forward that the speed of processes that take place in complex systems might be proportional to: $V \: e - (k \times I/T)$, where V = maximum speed and k = adjusting constant.

15. In several scientific forums there are now discussions on the behaviour of information and at last, a quantic theory of information has been drawn up. You can read something understand-

able in Poirier, H. (2005). *Aux limites de la matière, la realité n'est plus une certitude.* Science & Vie, 1057: 70-83. But the whole thing is at very early stages as yet.

16. It has been estimated that the human body replaces all its cells (except for neurons, ovules and a few others) every 7-10 years, but 98% of the atoms in the 26 different elements we are made up of (carbon, hydrogen, oxygen, nitrogen, iron, sodium, magnesium, etc.) are replaced every year. Taken from Thims, L. (2007). *Human Chemistry.* Morrisville, N.C: LuLu, 824 pp. In this regard, and if you have been married for quite some time with the same partner, as is my case, you'll have to accept you married someone else, molecularly speaking.

17. Apart from the candle, there are many systems that also have some features of life. Crystals, for instance, multiply in an appropriate solution and viruses can also replicate; they even have DNA, the molecule of life. However, viruses do not replicate by themselves and do not have their own metabolism (they are not autopoietic), so they need a living cell to multiply. They are loose programmes (RNA) -like computer viruses- which take control of the machinery of the infected cell. They do come from life but they are not life themselves; they can actually crystalize, like common salt.

18. The *walkabout* is or was an initiation ritual in which Australian aborigines would walk about the desert for six months to find themselves and become adults. Nowadays, all kind of Australians traditionally go about the world at least once in their life: a broad-spanning *walkabout.* I think western backpackers do something similar. At any rate, getting out into the world is a great idea, provided it is not done in a packaged tour. Being a tourist is not the same as travelling or being a wanderer.

19. Catalysts are substances that speed up or delay a chemical reaction without taking part in it. In autocatalysis, though, the substance in question induces and controls a chemical reaction on itself, thus altering its chemical structure without losing its autocatalytic properties in the process.

20. Smith, J. M. & Szathmáry, E. (1999). *The origins of life. From the birth of life to the origin of language.* Oxford: Oxford University Press, 179 pp.
21. Main milestones in the history of life on Earth (in years):
    4,600,000,000 Origin of the solar system and planet Earth
    4,500,000,000 Oldest rock (meteoritic origin) in Arizona.
    4,400,000,000 Gas emissions from Earth's mantle to the atmosphere.
    4,300,000,000 First mineral crystals (Australia). First continents?

    4,000,000,000 Formation of the Earth's crust, tectonic activity starts.
    3,900,000,000 Origin of bacteria (anaerobic, Archaebacteria)
    3,800,000,000 Calcium carbonate of biological origin (Greenland).
    3,600,000,000 First microfossils (stromatolites).
    3,500,000,000 Photosynthetic bacteria communities.
    3,300,000,000 Traces of gaseous oxygen (atmosphere and sediments)
    3,000,000,000 Appearance of metabolic types (H_2, H_2S, NH_3 and CH_4).

    2,500,000,000 Oxygen starts to accumulate; biogenic sediments.
    2,200,000,000 Profusion of plankton bacteria in the oceans (N fixing).
    2,100,000,000 Thickening of ozone layer (filters ultraviolet radiation).
    2,000,000,000 Plenty of O_2 in the atmosphere (aerobic organisms).
    1,800,000,000 Red colour iron sediments (rusted).
    1,700,000,000 First organisms with a nucleus, 'Protoctists'.
    1,600,000,000 Planktonic and benthonic organisms; mitochondria?
    1,500,000,000 Origin of mitosis, meiotic sex and programmed death.
    1,400,000,000 Cyanobacteria colonize Earth (desert crusts and soil).
    1,300,000,000 Diversification of multicellular algae: chloroplasts.
    900,000,000 Unidentified colonial protoctists in sandstone (Ediacara).
    900,000,000 Animals (with embryo and blastula), soft body.

    570,000,000 Hard animals (trilobites).
    500,000,000 Colonization of the emerged earth by algae and insects.
    440,000,000 Appearance of plants and fungi (earth groups).
    400,000,000 The first forests cover the Earth; plants with seeds.
    365,000,000 Marshy forests (Carboniferous).
    245,000,000 Age of Reptiles begins (Triassic, Jurassic and Cretacean).
    150,000,000 The supercontinent Pangea starts to break up.
    65,000,000 Extinction of dinosaurs and start of the Age of Mammals
    40,000,000 Appearance of plants with flowers and fruits. Primates.

23,000,000 Appearance of hominoids in Africa.
4,500,000 Human ancestors (biped hominid, *Australopithecus*).
2,500,000 Genus *Homo* (several species).
<100.000 Cultural evolution begins.

22. Reeves, H., Rosnay, J. d., Coppens, Y. & Simmonet, D. *La plus belle histoire du monde. Les secrets de nos origins*. This book deals with the history of the formation of the Universe and the Earth, the history of life and the history of humankind. Simmonet interviews the other authors: an astrophysicist, a biologist and a social anthropologist. A somewhat frenchified text that makes an easy instructive read.

23. The mutation rate in the mitochondrial DNA of the beetles I study, weevils of the genus *Laparocerus*, is 2.3% of the nucleotides per million years. This is generally the case in insects. Machado, A. *et al.* (2017). Phylogenetic analysis of the genus *Laparocerus* (Coleoptera, Curculionidae, Entiminae), with comments on colonization and diversification in Macaronesia. *Zookeys* 651: 1-77.

24. When the time comes for the cell to split, the genetic material (DNA) that is disperse in the nucleus is arranged, in more or less long chromosomes bodies. These are easy to see and count in a microscope. Their number varies from species to species and Science can now even count how many genes each of them are host to. In the case of *Homo sapiens*, there are about 30,000 genes, just 10,000 more than in ants. We share 98% of the genes with chimpanzees as we had a common ancestor about 4.4 million years ago (White *et al.* (2009). *Aridipithecus ramidus* and the paleobiology of early hominids. *Science* 326: 75-86).

25. Alleles are alternate forms of the same gene and there can be many different ones even though a cell may carry just two: one from paternal and the other from maternal origin. The differences between alleles—changes in their sequence of nucleotides—usually involve functional or structural variations (not as many). A familiar example is the gene that rules the colour of our eyes or our blood group.

26. The name of this principle is inspired in a passage in Saint Matthew's gospel 13:12 in which Jesus of Nazareth says: "For whosoever hath, to him shall be given, and he shall have more

abundance: but whosoever hath not, from him shall be taken away even that he hath". Don Ramón might just as well have worked harder on his inspiration or look for another metaphor because people now refers to this principle of dissymmetry on information as the Principle of St. Matthew of Margalef, without actually getting why an apostle is meddling in these scientific issues; or a still wilder idea: thinking that Margalef is a village in the Middle East.

27. I have an article where I approach life in more detail: Machado, A. (1999). *La vida en perspectiva* [Life in perspective]. Pp. 9-37 in: Fernández-Palacios J. M., *et al.* (ed.). *Ecología y cultura en Canarias.* La Laguna: Museo de la Ciencia y el Cosmos.

28. Gleick, J. *Chaos: Making a New Science* The theory of chaos is widely known as the "butterfly effect", an expression coined by an American meteorologist and mathematician, Edward Lorenz, while speaking about the climate as a chaotic system in which it is not possible to accurately determine what would happen when one part of the system, however small, is modified, he quoted a Chinese proverb: "a butterfly's flapping wings can cause a Tsunami at the other side of the world."

29. The Think Global School is a very original private school created by New Zealand millionaire Joan McPike. It focuses on the education years prior to University with the aim of training future adults and world leaders who are open-minded and tolerant. Curricula for change. The singular thing about this educational experiment is that teachers and students—some forty people—change country every term. So, it's a nomad school, with no fixed headquarters and unheard of. The nicest thing is you learn while travelling. In whichever country they are, they mingle with students from local schools, co-operate in community work, visit interesting places, and learn about their traditions. In short, they are impregnated with the culture of the country, until it's time to move on. And this runs parallel with the basic subjects they learn in combination with carrying out projects. Classes are in English and they also learn Mandarin and Spanish. The school has been working for a few years, has been to twelve countries and the ratio student teacher is 4:1.

30. Wagensberg, J. (2017). *Teoría de la creatividad* [Theory of creativity]. Barcelona: Tusquets Eds., Colección Metatemas, 288 pp.
31. Barto E. K., Weidenhamer J. D., Cipollini D., Rillig M. C. (2012). Fungal superhighways: do common mycorrhizal networks enhance below ground communication? *Trends in Plant Science* 17(11): 633-637.
32. Hölldobler, B. & Wilson, E. O. (1995). *Journey to the Ants: A Story of Scientific Exploration.* Cambridge: Harvard University Press, 304 pp.
33. The title of the short video is *Chimps and Tools*. You can watch it on http://video.nationalgeographic.com/video/chimp_tools. The images are better than the text.
34. *Cambridge Declaration on Consciousness* (2012): "The absence of a neocortex does not appear to preclude an organism from experiencing affective states. Convergent evidence indicates that non-human animals have the neuroanatomical, neurochemical, and neurophysiological substrates of conscious states along with the capacity to exhibit intentional behaviours. Consequently, the weight of evidence indicates that humans are not unique in possessing the neurological substrates that generate consciousness. Nonhuman animals, including all mammals and birds, and many other creatures, including octopuses, also possess these neurological substrates."
35. Human babies have the larynx in the same position as the other mammals and can breathe while sucking. Later, when they are more than two years old, the larynx descends down the oesophagus, enabling the production of the "base" sound that allows speech. You can read further in Arsuaga, J. L. & Martínez, I. (2005). *The Chosen Species: The Long March of Human Evolution.* New York: John Wiley & Sons, 298 pp.
36. In the early days, the atmosphere in our planet was white and was basically made up of carbon dioxide, ammoniac, molecular nitrogen and methane. Days used to be 10 hours long and years lasted for 876 days. Things change.
37. Going over my publications, I have found four where I deal with the psychosphere, defined as the layer on the Earth where thinking matter operates, and I discuss what this involves:

Machado, A. (2001). *De la biosfera a la psicosfera* [From the Biosphere to the Psychosphere]. Pp. 21-50 in: Marcos C., et al. (ed.). Gestión y ordenación del medio ambiente natural. Murcia: Universidad de Murcia, Servicio de Publicaciones.

Machado, A. (2006). *La psicosfera. ¿Necesitamos una nueva Ecología?* / The Psychosphere. Do we need a new Ecology?. Taro de Tahiche: Fundación César Manrique, 99 pp (bilingual)

Machado, A. (2006). *El rumbo del arca, entre la biosfera y la psicosfera* [The Course of the Ark, between the Biosphere and the Psychosphere]. Pp. 11-19 in: Mayol J. & Viada C. (ed.). El rumbo del arca. Palma de Mallorca: Conselleria de Medi Ambient del Govern de les Illes Balears.

Machado, A. (2015). The relevance of Infodynamics: From the Biosphere to the Psychosphere. *Eruditio, World Academy of Art and Science,* 1 (6): 107-109.

In them I include the following chart (with slight changes):

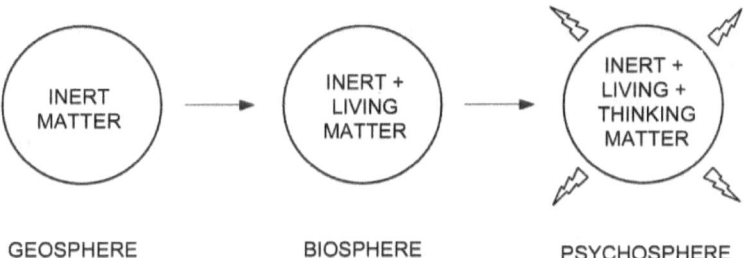

38. Peñuelas, J., 1988. De la biosfera a la antroposfera. Una introducción a la Ecología [From the Biosphere to the Anthroposphere. An introduction to Ecology]. Barcelona: Editorial Barcanova, 287 pp.
39. The genus *Homo*, which belongs to the order of Primates, is characterized by walking on its two hind feet which has prehensile toes and the first one is lined up with the rest. The skull is perfectly vertical with regard to the spine and has a volume over 700 cm^3. Their level of intelligence 5 includes the capacity to create tools and keep them for future use.
40. In the proteins that make up the living organisms we know, only 20 amino acids take part, which are very few in relation to the possibilities. This is a universal characteristic that champions a

single common origin. There is no reason why other forms of life based on water and carbon, or even on DNA, should use precisely these 20 amino acids. That is how it started, like one more contingency, and how it has kept, just like the chirality of the molecules: they are all left-handed.

41. The human brain is estimated to be made up of 100,000 million neurons, each of which may be connected to 1,000 or 10,000 other neurons, forming an absolutely colossal neuronal network. And there are ten glia cells per neuron. In fact, regarding cognitive capacity, the total number of neurons seems to be more important than the volume of the brain in relation to the body. In our species, neurons are very compact. Herculano-Heuzel, S. (2009). The human brain in numbers. A linearly scaled-up primate brain. *Frontiers in Human Neuroscience.* 3 (31): 1-11.

42. The idea of the existence of morphogenetic fields and morphic resonance comes from the 1920s (A. Gurwitsch, P. Weiss, etc.) although it is usually put down to Rupert Sheldrake, who developed it in depth in his book *Morphic resonance: the nature of formative causation* (Park Street Press, 2009). His postulate is that every object—living organisms, crystals, molecules—have a morphic field that, regardless of space and time, come into resonance with other similar fields and they influence each other on their respective constitutive patterns (formative causation hypothesis); it links past and future. In certain aspects, it reminded me of the *psi* factor—of a more powerful magic—which Ervin Laszlo presented in *The creative cosmos* (1993).

43. Part of the parental information that is conveyed to offspring apart from the nuclear DNA, resides in the cytoplasm of gametes (sexual cells), especially in ovules that, among other possible factors, carries organelles like mitochondria and chloroplasts (with their own membrane and DNA). In addition, at least in mammals, there are genomic processes that consist of reactions —like DNA methylation—that are produced during the formation of gametes and have later effects by activating or suppressing the expression of certain genes. This explains how descendants receive some experiences that marked their parents' lives, like phobia to specific smells. This type of inheritance, which

does not affect the nucleotide sequenced in the DNA, is called epigenetic and is being intensely studied. Obviously, its "marks" are also subject to natural selection.

44. TED (www.ted.com) is a non-profit organization that fosters the spreading of ideas, usually by means of intense short talks, about 18 minutes long, given by specialists. It started in 1984 with a conference on Technology, Entertainment and Design (hence the acronym TED) and it now includes any kind of topic of social interest. They are a very good alternative to so much trash TV and I highly recommend them.

45. At home I determined the exact wingspan of this male *Attacus atlas* as I had made marks on a piece of paper taking advantage of it being calmly alighted: it measures 24.6 cm. I checked it out on the internet and the species can be up to 25-30 cm, especially the females, which are bigger and stronger. However, I also read there is another species in Brazil, *Thysania agrippina* (the white devil) with a longer wingspan (31 cm), although it has less wing surface.

46. Biomass is a concept used in Ecology to refer to organic matter that makes up a living individual, a part, or the whole population of an ecosystem. It is usually measured in dry weight (after removing all the water), and in the last two cases it is expressed per area or volume unit. A standard man weighing 75 kg has a biomass of some 22.5 kg (dry weight). On the other hand, dead organic matter in an ecosystem is called necromass. The term biomass has recently become popular in the field of alternative energy but it does not mean exactly the same.

47. Until recently, it was accepted that *Homo sapiens* and *Pan troglodytes* (chimpanzee) shared a common ancestor 4.4 million years ago in Ethiopia, but recent genetic studies that take into account the mutation variation rate in primates, place this divergence far earlier, 12.1 million years ago (Moorjani, P., Amorim, C. E. G., Arndt, P. F & Przeworski, M. 2016., Variation in the molecular clock of primates. *Proceedings of the National Academy of Sciences of the United States of America*, 113 (38): 10607-10612).

48. Edward O. Wilson, in his many works, points out that individuals belonging to a social species are subject to two natural selection

processes which are, in a way, antagonistic: the selection of achievements that favour the individual him/herself (which happens in all the species) and the selection of achievements that favour the group (altruism, cohesion mechanisms, etc.). This means a certain balance must be kept for the species to succeed and progress as a whole. In human societies, currently subject to cultural evolution, extreme behaviours that favour the individual alone (gluttony avarice, lust, etc.) are valued negatively (sins); those behaviours favouring the group (mercy, justice, goodness, compassion, etc.) are considered virtues and positive to the group. This is a subtle example of selective cultural pressure via morality.

49. Kozlowski, J. R. (2016). Closed loop brain model of neocortical information based exchange. *Frontiers in Neuroanatomy* 10(3) PMID 26834573. The study is from the IBM research division, where they work with a neuronal tissue simulator.

50. Juan Benet's text is from the novel *Volverás a Región* and reads as follows:

"I believe the life of man is marked by three ages: the first is the age of impulse, in which our motivation and what we care about needs no justification; on the contrary, we are attracted to everything -a woman, a profession, a place to live in- thanks to an impulsive intuition that never compares; everything is so obvious that it is good in itself and the only thing that matters is the ability to obtain it. In the second age, everything we chose in the first phase has usually worn out, it is no longer good in itself and needs to be justified, something that the reasonable man does willingly, with the help of his heart, obviously; it is maturity, it is the time when, in order to come through unscathed from comparisons and the contradictory possibilities offered by everything around him, man makes that intellectual effort thanks to which the course chosen by instinct is later justified by reflection. In the third age, not only have the motivations chosen in the first age worn out and are no longer valid but also the reasons he underpinned his behaviour with in the second age. It is alienation, the rejection of everything his life has been, for which he no longer finds motivation or excuse. In order to be

able to live at peace, one must refuse to enter that third phase; however unnatural it may seem, he must make an effort and willingly remain in the second age; otherwise he will be adrift".

51. The Zen Namkhan Boutique Resort is situated below the tropic of Cancer, in parallel 19° 51' North latitude, the same one that goes half way between the Canary Islands and the Cape Verde archipelago or across Guantanamo, on the island of Cuba.
52. During the Vietnam War, from 1964 to 1973, when the country had some two million inhabitants, the United States dropped on Laos 4 million big bombs and some 270 million mini bombs contained in cluster bombs. Of this, it is estimated that 400,000 and 80 million respectively, did not explode properly. Data taken from Allman, T.D. (2015). Life after the bombs. *National Geographic*, 228 (2): 106-121.
53. In the end, I did not deal with the topic of tourism as it deserves but if you are concerned about it, I have an article on the situation I saw in Vietnam at the time the country was starting to open to tourism. Machado, A. (2004). *Turistas en Vietnam ¿bendición o condena?* [Tourists in Vietnam: blessing or bane?]. *Viajar,* 174: 26-31. You can find it in PDF format in the section 'Production/Tourism' on my website: www.antoniomachado.net
54. Moffett, M. W. (2011). *Adventures among ants. A global safari with a cast of trillions.* Berkeley: University of California Press, 288 pp. The same year, the *Scientific American* magazine published an extract entitled *Battles among ants resemble human warfare* in their December issue (pp. 84-89). It makes an entertaining read and conveys the essence of the book.
55. The Corbett tiger (*Panthera tigris corbetti*) is a subspecies of tiger that lives in Indochina, like the Bengal tiger (*Panthera tigris tigris*), another subspecies from India. Current population is estimated to be around a thousand individuals. There are countries, like Cambodia, where it was declared extinct in 2016. Programmes to recover them are being carried out in Vietnam. The main problem is that they are killed out of superstition, not because they eat people. Every single part of its body—including faeces—are used to prepare healing balms, making the most of

the animal strength they evoke and their mystic power. The most expensive part is their penis; pretty easy to guess why.

56. You might not know that when bees attack and sting you, they are actually sacrificing their lives to defend their group, the beehive. Their sting can only be used once: it is stuck on your skin and with it go part of its entrails, so it dies shortly after. In contrast, wasps sting you and inject their poison as if it were a syringe. They then live on just as happily.

57. One of the engineers of the company which set up the cremation ovens in Auschwitz, developed other improved models for the camps of Buchenwald, Dachau, Mauthausen, Gusen and Auschwitz too. He never understood why he wasn't congratulated on having created a superefficient system to burn corpses using the fuel of corpses themselves, thus saving on coal or oil.

58. In Biology, species are named in Latin by means of a binomial (e.g. *Elephas maximus* for the Asian elephant) and subspecies by a trinomial (e.g. *Elephas maximus indius*, the Indian elephant). In the early 1920s my Martian friend could have named the traditional human races subspecies thus: *Homo sapiens oceanicus* for Australians, *Homo sapiens luteolus* ("luteo" = yellow) for Asians, *Homo sapiens pallidus* for whites, *Homo sapiens rufescens* for the redskins, etc., and the African subspecies, which is where everything started, would be designated *Homo sapiens sapiens*. All these subspecies would make up the species *Homo sapiens*.

59. The concept of ecological niche is rather complex. It refers both to the task or function the species has in its ecosystem and to where and when it carries it out. Its ecological place, let's say. For instance, the wild cat is a medium-sized carnivorous predator of preferential night habits. It hunts at ground level or on trees across a large territory, sleeps and breeds in protected shelters, changing places if it perceives risk. That is what the ancestors of Meggy's, which is still by my side, would do. I'm afraid her niche boils down to the kitchen, the armchair, and meowing to get food.

60. Ways species are built up by splitting from an original one. Letters A, B, and C stand for different species; subscripts in A are

subspecies, the thick vertical line stands for a physical barrier (rivers, mountains, inlets, etc.)

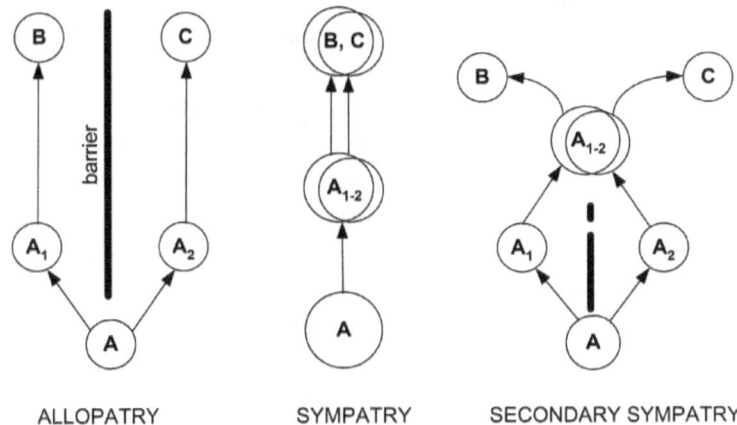

ALLOPATRY   SYMPATRY   SECONDARY SYMPATRY

61. For those who read Spanish, I eagerly recommend the first chapters of a book by Fernando Savater, where he offers an excellent explanation of the concept of citizenship. ¡*No te prives! En defensa de la ciudadanía* [Don't do without it! A defence of citizenship]. Editorial Ariel (2014), 224 pp.
62. The term panmixia (from Greek *pan* all, and *myctós* mixed) is used in Genetics and Biogeography to refer to the situation in which every individual in a given population has the same probabilities of mating with another of the opposite sex. They can all potentially mate, without distinction. There is a brief documentary uploaded by CISV (International Building Global Friendship) in YouTube, entitled *Momondo – The DNA Journey*, which shows the unsuspected levels of ADN mixtures from different sources which is already present in humans. It is quite illustrative to confirm the trend.
63. The resilience of a system is its capacity to absorb impacts and return to its natural condition, to normality.
64. *Ratatouille* is the title of a funny film by Pixar Studios (2007) which tells the story of a rat, Rémy, which is determined to become a chef. He is very bold and sneaks in a very posh French restaurant, where he'll be helped by a boy who turns out to be the hidden son of the owner. The evil one in the story is a gas-

tronomy critic who's feared by everyone because he's very elitist and demanding. But in the end, when he tries the first dish, a ratatouille, which has been secretly cooked by Rémy, he's impressed and has a flashback that takes him back to his tender childhood, to the ratatouille his mum used to cook for him. All his stiffness and sternness crumbles down. The term ratatouille is nicer than the word flashback and I believe it's not only used in this sense at home, and we understand each other. *Ratatouille* was awarded an Oscar for best animation picture.

65. Extract of one of the versions of the letter attributed to Chief Seattle, from the Suwamish tribe, to the president of the United States, Franklin Pierce (1864). The message is fake, but I didn't know that when I was a student. It reads: *"The Great Chief in Washington sends word that he wishes to buy our land [...] How can you buy or sell the sky – the warmth of the land? The idea is strange to us. [...] Every part of this earth is sacred to my people. Every shining pine needle, every sandy shore, every mist in the dark woods, every clearing, and every buzzing insect is holy in the memory and experience of my people. The sap that runs through the trees carries the memories of the red-skinned man [...] The rivers are our sisters and quench our thirst [...] The white man [...] treats his mother, the earth, and his brother the heavens, as if they were things that could be bought, looted and sold, as though they were lambs and glass beads. His insatiable hunger will consume the earth and leave behind a desert [...] You must teach your children what we have taught ours: that the earth is our mother. Everything that affects the earth affects the sons of the earth. [...] The earth does not belong to man. Man belongs to the earth. We know this. All things are bound up in each other like the blood that binds the family. Man has not woven the fabric of life: he is just a thread in it. Everything he does to this fabric he does to himself."*

66. Gaia is the Greek goddess of Earth, she represents mother Earth, nature. Her name was used, following the suggestion of writer William Golding, to call the Gaia Hypothesis put forward by the chemist James Lovelock in 1979. It says that planet Earth, with its biosphere, works like a conscious superorganism that regulates

itself to keep in balance. Lovelock himself gave up the idea because it is teleological but thousands of Gaians did not do the same and keep on following this alluring view. More so after the unswerving campaign made by Lynn Margulis, the biologist who discovered symbiogenesis. She is a great scientist who is never discouraged, but she's also one of the most stubborn persons I've met.

67. To shoot an enemy ship with a cannon you have to measure distance, wind speed and other parameters which, together with the amount of gunpowder and the size of the cannonball, allow to calculate the inclination of the cannon required before shooting (with so much calculation, the ship might just sail past). Cannon shooters in the past used to employ a technical approach based on experience: the "bracketing". They would fire a long shot and then a short one, see where the shells fell and then decide proportionally on the right inclination of the cannon. The third shot usually hit the ship bang on.
68. The principle of cautiousness or precaution suggests adopting of protective measures against a threat before there is an irreparable deterioration of the environment, even if there is no scientific certainty about the cause and effect of the said threat.
69. There are five ways of approaching nature conservation: two are preventive: (1) planning and (2) previous studies on environmental impact; then, (3) sustainable management of species, not allowing their exploitation beyond their capacity to renovate; there is also (4), preservation, that is, total protection, nothing can be touched and finally (5) recovery, which is an attempt to recover damaged nature. In the past, environmentalists and conservationists focused solely on preservation (4) but they've slowly opened to other options. Nonetheless, many of them have not evolved, including some scientists who are also environmentalists.
70. I use a Spanish transcription of the English term paper which is used, in some academic areas, to refer to a scientific article. It's a sad and affected Anglicism that is getting into Spanish. It's not out of snobbism but rather a visible mark of the inferiority complex of our science compared to the Anglo-Saxon. The same

goes for the habit of using capital letters in nouns for the titles of articles.

71. Interpretive planning refers to the techniques developed to interpret the language of nature, especially in natural parks, so that visitors can enjoy the natural environment better, as they get a fuller understanding of it. It's something quite empathic and ludic and I like relating it to environmental seduction rather than to formal environmental education, which is often focused on the mere instruction about the environment, at least in Spain.

72. García de Enterría, E. (1974). *La lucha contra las inmunidades del poder* [The Struggle against the Immunity of Power]. Madrid: Cuadernos Civitas, 99 pp. This little book displays great democratic lucidity and I wholeheartedly recommend it to every civil servant or public official that has reached positions where they hold certain responsibilities in the government or in public administration, provided that they read Spanish, of course.

73. It is commonly accepted that in the history of the Earth there have been five major extinctions due to different natural phenomena (glaciations, meteors, etc.), although the first great extinction is not included: that caused by the first bacteria that released oxygen into the environment, poisoning their congeners (no fossils remain). In this catastrophic context, the sixth great extinction is considered to be underway and it's due to *Homo sapiens*. It's estimated that in the last 100 years we've caused the extinction of as many species as in 10,000 years in natural conditions. See Leakey, R. E. & Lewin, R. (1997). *The Sixth Extinction: Patterns of Life and the Future of Humankind.* Woodson: Anchor Books, 271 pp.

74. Albedo is the percentage of radiation a surface reflects from what it gets. The average albedo of solar radiation on Earth is of 37-39%. Albedo increases in the large deforested surfaces as these have gone from dark to lighter colours, affecting climate.

75. Halting the loss of biodiversity by 2010 –and beyond. Sustaining ecosystem services for human well-being. Brussels: Commission of the European Communities COM (2006) 216 final, 15 pp.

76. The sentence is Margalef's. From the point of view of chemistry, biomass represents a reduced state of carbon, with its many hy-

drogen links. On "burning" biomass, carbon ends up combining again with oxygen (CO_2), that is, oxidized, which was its original condition.

77. The most energetic thing on our planet and the driving force behind the sudden great changes in the biosphere is Layer D, which is located between the outer core and the mantle of the Earth's lithosphere. Its dynamics cause intermittent convections in the mantle, super-plumes that reach the crust, bulging that lead to drastic sea regressions or transgressions, displacement of continents, etc. At its time scale, it's very active and its temperature is above the minimum needed to warrant chaotic behaviour, although the changes it provokes have always been tried to be labelled periodical (glaciations, extinctions, climate). The same is sought in the influence of the sun or the presumed cosmic cycles.

78. The level of energy managed by the climate ranges from 210,000–240,000 kw/km². The general level that can be put down to our species is around 16 kw/km², equivalent to the wind's maximum energy on the ground or the sea. The average in the United States is 318 kw/km². Only in places such as Manhattan peaks reach 630,000 kw/km². (Margalef, R. 1997. *Our Biosphere*).

79. The proposal for a "Universal Declaration of Democracy" was promoted by Federico Mayor Zaragoza when he was General Secretary of UNESCO. The following is article 1: "*Democracy is a political, economic, social, cultural and international regime, based on the respect for human being, the supremacy and independence of justice and law, as well as on the possibility for any individual to participate in the life and development of society, in freedom and peace and in a favourable natural and cultural environment, being always fully conscious of the equal dignity and interdependence of the human beings*". And article n° 29, which reads as follows: "*All human beings have the duty to respect and defend democracy and peace in their various fields of operation: political, economic, social, cultural and international. They shall in no circumstances exercise or defend their rights in ways contrary to the aims and principles of United Nations*". Taken from Mayor Zaragoza, F. (2013). For a universal

declaration of democracy. *Eruditio* 1 (1): 1-10. The underlining in the word "duty" is mine because there being a *Declaration of Human Rights* by the United Nations, I have always missed its counterpart, a "Declaration of Human Duties".

80. Wagensberg, J. (2015). Método Wagensberg, el 'gozo palanca'. [The Wagensberg Method: the 'lever joy']. Mètode - Revista de difusión de la Investigación, 85: 11.

81. *WhatsApp* is an application for smartphones that sends and receives messages through the internet. You can create groups and share images, music and all kind of nonsense. It is overwhelmingly successful and in 2016 it had more than 1,000 million users who can whatsapp for as long as they like, till they reach affective castration.

82. At a school-restaurant in Luang Prabang I saw very interesting indications in the menu. You could choose: European spicy = 1 chilli, Laotian medium spicy = 3 chillies or Laotian really spicy = 5 chillies. These high levels of pleasure can only be enjoyed with dignity after thorough training.

83. Magrassi, L. & Leto, K., Rossi, F. (2013). Lifespan of neurons is uncoupled from organismal lifespan. *Proceedings of the National Academy of Science U S A.* 110 (11): 4374-4379.

84.

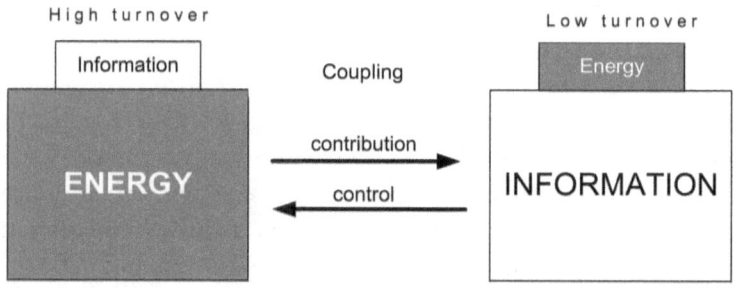

85. Among the primates, *Homo sapiens* stands out for its sexuality: it has the longest and thickest penis, spends more time courting, copulation lasts longer (4-7 minutes); the female ovulation is hidden to the male, so she's always receptive; her breasts are big outside the breastfeeding period (an attractive feature) and her

clitoris is an organ exclusively for pleasure, therefore orgasms are not limited to being on heat. All these adaptations seem to encourage the link in couples, but there are all kinds of interpretations.

86. Mathematics is not a science in as far as it does not study physical reality. It is a language, an accurate language used by many sciences as a tool.
87. This mental game can also be applied to the sea, replacing the water dial with a light dial; that is, temperature, light and nutrients. With the three dials to the full, we would get a coral reef. Someone could develop a simple simulation game with this idea and show the main biome on the planet by using these three dials.
88. http://www.antoniomachado.net
89. Wikipedia definition of *ikigai*: "Japanese concept that means a reason for being. Everyone, according to Japanese culture, has an *ikigai*. Finding it requires a deep and often lengthy search of self. Such a search is important to the cultural belief that discovering one's *ikigai* brings satisfaction and meaning of life".
90. I have depicted this attitude in a short tale *The King's Dwarf*, which I copy below:

Moses was devoted to carving wood and he took to making garden dwarves which he would then colour and provide with some distinguishing feature, so that every flower bed or balcony inhabited by one of his carefully-made creatures would get a stamp of its own. All the fortunate owners had this possession in great esteem.

One sunny morning, filled with the singing of the birds, Moses heard how the clanking of horse hooves pulling a carriage stopped outside his workshop. He put down his burin, brushed away the sawdust from his apron and waited, carefully watching the shadows that filtered under the door. It was not long before a well-shaven and groomed man in rich garments and courtesan manners appeared and spoke thus:

"Are you Moses, the dwarf-maker?"

"I am; at your service", our man answered with a brief bow of his head in acknowledgment of the gentlemanly manners he had perceived in his visitor.

"I am the King's chamberlain". The fame of your dwarves has reached the Palace and His Majesty wants you to make one for his private courtyard.

"It could not be easier, my lord", and he turned around and took one of the dwarves which were lined up on a shelf behind him. He offered it to the distinguish visitor with a beatific smile.

This caused an equally beatific smile in the chamberlain and, as if he were explaining the most obvious thing in the world to a schoolboy, with wide-opened eyes, benevolently pointed out:

"My dear master Moses, I might not have made myself clear. We are talking about a dwarf for the King...

The craftsman was quick to answer:

"Excellency, all my dwarves are made as if they were for the King".

91. I do not think it is still there, but many years ago, when I used to travel frequently to Europe and before airports turned into shopping centres, there was, in the middle of the main hall at Paris Charles De Gaulle, a fancy kiosk that offered French perfume and fashion jewellery. On the top there was a great notice: «*Faites-vous pardonner vos sorties*» [Help forgive your outings]. I thought it was a funny idea.

92. A very active physicist with no formal academic qualifications, Nassim Haramein, has postulated a unified memory-space network theory that would help to explain the fundamental nature of space, time, energy, and matter according to geometric and informational interactions, encompassing from quantic scales to larger ones, from cosmogenesis to the conscience of the universe itself. He puts forward the idea that the void is not such but that is full of energy that unites the whole Universe. His latest publication is very recent and in it, he claims to have found the philosopher's stone of Physics: a unified theory of the universe (Haramein, N., Brown, W. D. & Baker, A. V. The Unified Spacememory Network: from cosmogenesis to consciousness. *NeuroQuantology* 2016 (14): 1-15). There will be a lot of talk

about it, as it has happened with everything this controversial character has dealt with, including the relationships of aliens with the Earth.

93. Having rounded off my manuscript, I read Wilson's book. I found it rather astonishing because I don't think it keeps up to its title. I'm not sure whether other readers will actually understand what he thinks about human existence and what declaring himself "existential conservative" involves. As for the rest, he once again deals with the evolution of our species and speculates on its plausible future. Except for this last new aspect, he tackles recurring topics in his writings, like knowledge unicity (merger of science and arts) and, above all, the selective factors that affect *Homo sapiens* as a social animal, dedicating a long chunk of the text to rebutting the "Inclusive fitness theory" in a repetitive, obsessive way, perhaps making a last effort to reaffirm his sociobiological theory within the framework of evolutionary biology.

[Content Guide]

Preface 9
Day One 15
    [Types of knowledge] 16
    [Organization of the book] 18
    [The bulbul and the watch] 20
    [Systems theory] 21
Day Two 25
    [On big and small] 26
    [The moviola of time] 28
    [The Monsoon] 29
    [Types of information] 31
    [Laotian Funeral] 34
Day Three 39
    [Complex adaptive systems] 40
    [Inert Matter] 42
    [Living Matter] 42
    [Mitochondrion and chloroplast] 45
    [Oikeiosis] 47

Day Four 49
   [Autocatalysis] 50
   [Emerging properties] 51
   [The origin of life] 52
   [Mutations, sex and death] 54
   [Contingencies] 57
   [St. Matthew's principle] 58

Day Five 61
   [The Namkhan River] 62
   [The biosphere] 63
   [Theory of chaos] 64
   [Rising complexity] 65
   [The meaning of life] 66
   [Hamza, the Moroccan] 67

Day Six 71
   [Rising communication] 71
   [Levels of intelligence] 72
   [Thinking matter] 76
   [The psychosphere] 78
   [Life - mind differences] 79
   [Souls and gods] 80
   [Ideas and creativity] 80
   [Cultural evolution] 81
   [Will and finalism] 82
   [Information ecology] 83

Day Seven 85
   [Human achievements] 87
   [The meaning of the mind] 89
   [The individual and the mind] 91
   [The heart of darkness] 93
   [The light of understanding] 95

Day Eight 99
[Ban Suan Village] 100
[Itinerant agriculture] 101
[Tad Sae Waterfalls] 102
[Instincts] 104
[Tribalism and civilization] 105
[Religions and superstitions] 107

Day Nine 111
[Reason and barbarism] 112
[Geographical speciation] 114
[Races and subspecies] 115
[Secondary sympatry] 116
[Cultural racism and globalization] 119
[The birds which sing best] 121

Day Ten 123
[Nature conservation] 124
[Reason and legitimacy] 128
[Vocation and ethics] 129
[Environmentalism and ecofascism] 130
[Endangered planet?] 131
[Impact of the mind] 133
[Energy inputs] 135
[Climate change] 136

Day Eleven 139
[Ecological footprint] 140
[Ugliness and harmony] 141
[The evils of plenty] 143
[Doses and resilience] 145
[Infoxication] 146
[Smartphones] 148
[Advertising] 150

   [Technology and its toll] 152
Day Twelve 155
   [Ecosystems, youth and maturity] 156
   [Coupling and exploitation] 158
   [Pareto and St. Matthew] 158
   [Human feelings] 160
   [Hard and soft models] 162
   [Money and territory] 162
   [The market] 165
   [Lies and deceit] 166
Day Thirteen 169
   [Happiness] 170
   [Luang Prabang] 170
   [Humanine] 173
   [Ikigai] 175
   [My case] 177
   [Transcendence] 178
Day Fourteen and Last 181
   [Parade of monks] 181
   [Morning Market] 182
   [Physical recapitulation] 185
   [Bangkok Airport] 187
Epilogue 189
Acknowledgements 191
Appendix 193
Final Notes 198

Dr. Machado (Madrid, 1953) has been a lecturer in Ecology at the University of La Laguna (Tenerife, Canary islands), superintendent of Teide National Park, adviser on environmental policies for the Presidency of the Spanish Government, Regional Councillor of the IUCN (International Union for Conservation of Nature) and President of the ECNC (European Centre for Nature Conservation). He has also worked as independent consultant for several international organizations and foreign co-operation programmes. He is currently the head of a public foundation in the Canary Islands (Granadilla Environmental Observatory) and researches privately in Entomology. He is also editor-in-chief of the Journal for Nature Conservation (Elsevier group) and is a full member of the World Academy of Arts and Science and of the Canarian Academy of Language. In addition to his vast scientific writings, he has also written more general books like *Ecología, medio ambiente y desarrollo turístico en Canarias* (1990), *T. Vernon Wollaston (1822-1878). Un entomólogo en la Macaronesia* (2006), *The Psychosphere. ¿Do we need a new Ecology* (2006) or *28-D Enclave de humor* (2012), the latter with filmmaker Santiago Ríos.

www.ingramcontent.com/pod-product-compliance
Lightning Source LLC
Chambersburg PA
CBHW030618220526
45463CB00004B/1340